INTERNATIONAL INSTITUTE OF PHYSICS
founded by Ernest Solvay

PROCEEDINGS OF THE
FOURTEENTH CONFERENCE ON PHYSICS
AT THE UNIVERSITY OF BRUSSELS
OCTOBER 1967

International Inst.-

Fundamental Problems in Elementary Particle Physics

1968
INTERSCIENCE PUBLISHERS
a division of John Wiley & Sons, Ltd.
LONDON · NEW YORK · SYDNEY

First published 1968 by John Wiley and Sons, Ltd.
All rights reserved. No part of this book may be
reproduced by any means, nor transmitted, nor
translated into any machine language without the
written permission of the publisher.

Library of Congress Catalog Card No. 68-8866

SBN 470 81256 7

Printed in Northern Ireland at The Universities Press, Belfast

QUATORZIEME CONSEIL INTERNATIONAL DE PHYSIQUE

Commission Administrative

Président:	F. LEBLANC, Président du Conseil d'Administration de l'Université Libre de Bruxelles.
Membres:	P. DE GROOTE, Président honoraire du Conseil d'Administration de l'Université Libre de Bruxelles.
	E. J. SOLVAY, Membre du Conseil d'Administration de l'Université Libre de Bruxelles.
	J. SOLVAY, Membre du Conseil d'Administration de l'Université Libre de Bruxelles.
	L. DEBROUCKERE, Professeur à l'Université Libre de Bruxelles.
Directeur:	I. PRIGOGINE, Professeur à l'Université Libre de Bruxelles.
Secrétaire:	R. LECLERCQ, Secrétaire général de l'Université Libre de Bruxelles.

Comité Scientifique

E. AMALDI	Universita degli Studi, Roma, Italia.
SIR LAWRENCE BRAGG, F.R.S.	The Royal Institution, London, England. (excusé).
C. J. GORTER	Kamerlingh Onnes Laboratorium, Leiden, Nederland.
W. HEISENBERG	Max-Planck Institut für Physik und Astrophysik, München, Deutsche Bundesrepubliek,
C. MØLLER (*President du XIVe Conseil International de Physique*)	Nordisk Institut for Teoretisk Atomfysik, København, Danmark.
F. PERRIN	Commissariat à l'Energie Atomique, Paris, France.
I. TAMM	P. N. Lebedev Physical Institute of the Academy of Sciences, Moscow, U.S.S.R. (excusé).
S. TOMONAGA	The Tokyo University of Education, Bunkyo-ku, Tokyo, Japan. (excusé).
J. GÉHÉNIAU (*Secrétaire*)	Université Libre de Bruxelles.

Membres Rapporteurs

G. F. CHEW	University of California, Berkeley, California, U.S.A.
H. DÜRR	Max-Planck Institut für Physik und Astrophysik, München, Deutsche Bundesrepubliek.
M. GELL-MANN	California Institute of Technology, Pasadena, California, U.S.A.
R. HAAG	Universität Hamburg, Institut für Theoretische Physik, Hamburg, Deutsche Bundesrepubliek.
W. HEISENBERG	Max-Planck Institut für Physik und Astrophysik, München, Deutsche Bundesrepubliek.
G. KÄLLÉN	Institute of Theoretical Physics, University of Lund, Sweden.
E. G. C. SUDARSHAN	Syracuse University, Syracuse, New York, U.S.A.
A. TAVKHELIDZE	Joint Institute for Nuclear Research, Dubna, U.S.S.R.

Membres Invités

S. L. ADLER	The Institute for Advanced Study, Princeton, New Jersey, U.S.A.
N. CABIBBO	CERN, 1211, Genève 23, Suisse.
B. FERRETTI	Academia Nazionale dei Lincei, Bologna, Italia.
M. FROISSART	Commissariat à l'Energie Atomique, Centre d'Etudes Nucléaires de Saclay, France.
S. FUBINI	Universita di Torino, Torino, Italia.
J. HAMILTON	Nordisk Institut for Teoretisk Atomfysik, København, Danmark.
R. HOFSTADTER	Stanford University, Stanford, California, U.S.A.
M. LÉVY	Université de Paris, Paris, France.
F. E. LOW	Massachusetts Institute of Technology, Cambridge, Massachusetts, U.S.A.
S. MANDELSTAM	University of California, Berkeley, California, U.S.A.
R. E. MARSHAK	The University of Rochester, Rochester, New York, U.S.A.
L. MICHEL	Institut des Hautes Etudes Scientifiques, Bures-sur-Yvette, France.

L. B. OKUN	Institute of Theoretical and Experimental Physics, Moscow, U.S.S.R.
R. OMNES	Centre de Recherches Nucléaires, Strasbourg Cronenbourg, France.
L. A. RADICATI	Scuola Normale Superiore, Pisa, Italia.
T. REGGE	Universita di Torino, Torino, Italia.
L. ROSENFELD	Nordisk Institut for Teoretisk Atomfysik, København, Danmark.
D. RUELLE	Institut des Hautes Etudes Scientifiques, Bures-sur-Yvette, France.
S. SAKATA	Nagoya University, Chikusa-ku, Nagoya, Japan.
J. SCHWINGER	Harvard University, Cambridge, Massachusetts, U.S.A.
H. UMEZAWA	University of Tokyo, Bunkyo-ku, Tokyo, Japan.
C. F. VON WEIZSÄCKER	Universität Hamburg, Hamburg, Deutsche Bundesrepubliek.
S. WEINBERG	Harvard University, Cambridge, Massachusetts, U.S.A.
A. S. WIGHTMAN	Princeton University, Princeton, New Jersey, U.S.A.
E. P. WIGNER	Princeton University, Princeton, New Jersey, U.S.A.
S. A. WOUTHUIJSEN	Universiteit van Amsterdam, Amsterdam, Nederland.

Membres Secrétaires

J. GÉHÉNIAU	Université Libre de Bruxelles.
F. HENIN	Université Libre de Bruxelles.
J. NAISSE	Université Libre de Bruxelles.
J. BIJTEBIER	Université Libre de Bruxelles.
P. CASTOLDI	Université Libre de Bruxelles.
M. EVRARD-MUSETTE	Université Libre de Bruxelles.
C. SCHOMBLOND	Université Libre de Bruxelles.
C. GEORGE	Université Libre de Bruxelles.

Membres Auditeurs

R. BROUT	Université Libre de Bruxelles.
M. CAHEN	Université Libre de Bruxelles.
F. CERULUS	Université de Louvain.
R. DEBEVER	Université Libre de Bruxelles.
F. ENGLERT	Université Libre de Bruxelles.
P. GLANSDORFF	Université Libre de Bruxelles.
R. HERMAN	General Motor Research Laboratories, Warren, Michigan, U.S.A.
C. JOACHAIN	Université Libre de Bruxelles.
M. LAMBERT	Université Libre de Bruxelles.
C. MANNEBACK	Académie Royale de Belgique.
V. MATHOT	Université Libre de Bruxelles.
J. PELSENEER	Université Libre de Bruxelles.
J. PHILIPPOT	Université Libre de Bruxelles.
J. REIGNIER	Université Libre de Bruxelles.
J. SACTON	Université Libre de Bruxelles.
J. SERPE	Université de Liège.
D. SPEISER	Université de Louvain.
G. THOMAES	Université Libre de Bruxelles.
M. VELARDE	Université Libre de Bruxelles.

CONTENTS

R. MØLLER
(*President*)   Homage to R. Oppenheimer . . xi

Reports and Discussions

H. P. DÜRR   Goldstone Theorem and Possible Applications to Elementary Particle Physics 1
Discussion on the report of H. P. Dürr   18

R. HAAG   Mathematical Aspects of Quantum Field Theory 23
Discussion on the report of R. Haag   32

G. KÄLLÉN   Different Approaches to Field Theory, Especially Quantum Electrodynamics   33
Discussion on the report of G. Källén   48

G. F. CHEW S-Matrix Theory with Regge Poles   65
Discussion on the report of G. F. Chew 87

E. C. G. SUDARSHAN   Indefinite Metric and Nonlocal Field Theories 97
Discussion on the report of E. C. G. Sudarshan 125

W. HEISENBERG   Report on the Present Situation in the Nonlinear Spinor Theory of Elementary Particles 129
Discussion on the report of W. Heisenberg 141

A. TAVKHELIDZE   Simplest Dynamic Models of Composite Particles 145
Discussion on the report of A. Tavkhelidze 153

I. PRIGOGINE	Dissipative Processes, Quantum States and Field Theory	155
	Discussion on the communication of I. Prigogine	196
General Discussion	. . . . . .	203
L. ROSENFELD	Some Concluding Remarks and Reminiscences	231
Discussion on the report of M. Gell-Mann	. . . .	235
Index	. . . . . .	245

HOMAGE TO R. OPPENHEIMER

In the middle of the preparations for the conference our president Robert Oppenheimer died on February 18th this year. Until about a week before he died he still hoped to be able to attend the meeting, the subject of which was particularly close to his heart.

With Oppenheimer's passing away international physics lost one of its leading figures and U.S.A. lost one of its great sons. His contributions to theoretical physics covered such a large field that it would be impossible, and in this circle also unnecessary, to refer to all of them here. Among his most important works I shall only mention

(1) his early contribution to the quantum theory of molecules,
(2) his recognition of the importance and his treatment of exchange effects in electron–electron scattering,
(3) his various contributions to the theory of positive and negative electrons and to quantum electrodynamics,
(4) his detailed account of the cascade mechanism in cosmic ray showers, and
(5) his work on the possibility of gravitational collapse of massive stars, which has become of renewed interest in connection with recent astrophysical discoveries.

Many of these investigations were performed in collaboration with his students, a consequence of his brilliant activity as teacher in theoretical physics.

During his lifetime he created two great schools of theoretical physics, at Berkeley and Pasadena in the thirties, and at Princeton after 1947. His guidance of the work of numerous students and the inspiration and enthusiasm he conveyed to them has perhaps been even more important for the development of American physics than his own published papers.

During World War II, Oppenheimer organized the Los Alamos laboratory and directed the work which led to the construction of the first atomic weapons. His enormous contribution to the security of his country was recognized by the award of the Medal of Merit in 1946 and made him a public figure in the years after the war. Therefore it came as a shock to him and to his many friends all over the world when his security clearance was withdrawn in 1954. The shattering experiences in this dark period of American politics partly ruined his health and may very well have been a contributory cause of his untimely death.

His subsequent rehabilitation at the end of 1963, when the Enrico Fermi prize was awarded him, actually *came too late*. When I shortly afterwards had the occasion to meet him, I got the impression that this also was his own opinion, for when I congratulated him and told him how pleased we were, he said after a moment of silence "yes, this award has at least done one good thing, it has made my friends happy".

Oppenheimer, although truly American, had a strong feeling of solidarity with Europe and a deep understanding of European problems and European culture. In his youth he spent a number of years in Europe (he got his doctors degree at Göttingen and was subsequently a research fellow at Leiden and Zürich) and on many occasions he expressed his indebtedness to European culture. Six years ago he readily accepted our presidency and we vividly recall the brilliant way in which he conducted our meetings. He will be sorely missed at this conference and at our future meetings.

PROFESSOR C. MØLLER
President of the XIVe Conseil International de Physique

GOLDSTONE THEOREM AND POSSIBLE APPLICATIONS TO ELEMENTARY PARTICLE PHYSICS

H. P. Dürr

Max-Planck-Institut für Physik und Astrophysik, München, Germany

The Goldstone theorem states that under certain conditions, which will be stated later, in a dynamical theory which is invariant under a particular symmetry group and where this symmetry group is broken by the ground-state, e.g. the vacuum state in a relativistic field theory, there must exist particles of mass zero or—in the nonrelativistic case—excitation modes, the energy of which tends to zero with increasing wavelength. This theorem is of great interest for elementary particle physics for essentially two reasons:

(1) There are a number of symmetry groups in elementary particle physics, like $SU(3)$ or even higher symmetries, which are not exactly realized in nature in the sense that there exist one-particle states or resonances grouped into multiplets which are supposed to transform approximately according to an irreducible representation of this group but which do not have exactly the same mass, and secondly that in the interaction of these particles and resonances the conservation laws, related to this symmetry group by Noether's theorem, are only approximately obeyed. On the other hand, these symmetry violations do not seem to be connected in a direct or an indirect way with the appearance of mass zero particles. Therefore, *if* one wants to interpret this symmetry violation as a consequence of an asymmetrical vacuum as proposed for $SU(3)$ by Baker and Glashow[1], rather than as an asymmetry of the underlying dynamical law, one has to find some means to *invalidate* the Goldstone theorem.

(2) There do exist in nature a number of massless particles: the photon, the neutrinos and, probably, the graviton. In a general dynamical theory it is rather difficult to obtain such particles as particular solutions, except by chance or if they are introduced from the beginning. In this context the Goldstone theorem, it appears, could provide an interesting way to enforce their existence, because, in fact, all the known massless particles do occur in connexion with symmetry violations, the photon with isospin violation, the neutrinos with parity violation, etc., and the graviton

probably with a violation of the Poincaré group[2]. Unfortunately, however, it turns out that all these mass zero particles do *not* have the symmetry properties, as explicitly stated by the Goldstone theorem in its present mathematical formulation. Therefore, in order to uphold this conjecture, the present predictions of the Goldstone theorem on the symmetry properties of the massless particle must be generalized.

Consequently the validity of the conjecture that the observed symmetry violations in elementary particle physics arise from an asymmetry of the vacuum state, will decisively depend on an invalidation or—if we exclude $SU(3)$ and possible higher symmetries as fundamental symmetries—on a generalization of the Goldstone theorem. In contrast to this, in non-relativistic dynamics we know many systems where the Goldstone theorem holds in its present form. The magnons in the ferromagnet, the phonons in liquids and crystals are, for example, Goldstone modes connected with an asymmetry of the groundstate[3].

There exist many general proofs of the Goldstone theorem today. The first proofs were given by Goldstone, Salam and Weinberg[4] and by Bludman and Klein[5]. A proof on a much more rigorous basis using only the algebra of observables was recently given by Kastler, Robinson, and Swieca, and Ezawa and Swieca[6].

Let me roughly sketch the proof of the Goldstone theorem in order to indicate the basic assumptions. Let us assume there is a certain symmetry transformation which leaves the dynamics invariant. Formally, this may be expressed by the forminvariance of a Lagrangian density or the forminvariance of an equation of motion and the quantization condition. As an example we may just use a simple gauge transformation to simplify the discussion. As a consequence of the invariance there exists, according to the Noether theorem, a conserved current:

$$\partial_\mu j^\mu(x) = 0 \tag{1}$$

and a time-independent hermitean operator

$$Q = \int d\sigma^\mu j_\mu(x) = \int_{x^0=t} d^3x \, j^0(x) \tag{2}$$

which serves as a generator of the unitary representation of the symmetry group in the state space. The symmetry is *broken* by the translational invariant vacuum state, if for some field operator $\phi(x)$, which is not invariant under this symmetry group, i.e. which transforms as $\phi(x) \to \phi'(x)$, the vacuum expectation value changes

$$\langle 0| \phi(x) |0\rangle = \langle 0| \phi(0) |0\rangle \neq \langle 0| \phi'(x) |0\rangle = \langle 0| \phi'(0) |0\rangle \tag{3}$$

or, expressed differently, if for an infinitesimal symmetry transformation $\sim \delta\lambda$ we have

$$\frac{1}{\delta\lambda} \langle 0| \delta_{\text{sym}} \phi(x) |0\rangle = -i \langle 0| [Q, \phi(x)] |0\rangle = C \neq 0 \tag{4}$$

(with C = constant). By introducing a complete set of intermediate states the latter may also be expressed as[7]

$$\sum_{0' \neq 0} 2 \mathscr{I}m \langle 0| Q |0'\rangle \langle 0'| \phi(x) |0\rangle = C \neq 0. \tag{5}$$

If we make use of the local form (2) of the generator one gets

$$-i \int d\sigma'^{\mu} \langle 0| [j_\mu(x'), \phi(x)] |0\rangle = C \neq 0 \tag{6}$$
(for all surfaces)

which leads to the local condition

$$\langle 0| [j_\mu(x'), \phi(x)] |0\rangle = \frac{\partial}{\partial z^\mu} f(z) \qquad (z = x - x') \tag{7}$$

with

$$-i \int d\sigma'^{\mu} \frac{\partial}{\partial z^\mu} f(z) = C.$$

Due to the locality requirement (vanishing of the commutator for space-like distances) $f(z)$ can be written as a superposition of causal functions $\Delta(z; m)$ with various masses m:

$$f(z) = iC \int dm^2 \rho(m^2) \Delta(z; m) \qquad \int dm^2 \rho(m^2) = 1. \tag{8}$$

As a consequence of the current conservation (1)

$$\partial_z^2 f(z) = 0 \quad \text{and hence} \quad \rho(m^2) = \delta(m^2) \tag{9}$$

i.e. $\phi(x)$ must contain matrix elements leading to massless particles from the vacuum. This is the content of the Goldstone theorem.

From Eqn. (5) one merely deduces that there exists in the theory other states $|0'\rangle$ different from the vacuum state $|0\rangle$ which have the same energy (and momentum) as the vacuum state since Q is a (time independent) symmetry operator. We call these states $|0'\rangle$ spurion states. They are created from the vacuum by $\phi(x)$. Relation (9) which has made use of the local structure of the symmetry operator and hence contains more information, reveals that the spurions, in fact, are merely the infrared limit ($\mathbf{p} \to 0$) of massless particles generated by $\phi(x)$, the Goldstone particles, which we will shortly call 'zerons'. Roughly speaking the zerons are localized spurions.

In this connexion Heisenberg[2] has emphasized that any localized spurion connected with a nonlocalized spurion could again be a possible 'zeron'. Hence the symmetry character of the Goldstone zeron should in general not be immediately identified with the symmetry character of the spurions which is usually done, but states a separate problem.

I wish now to remark on the various steps of the rough (and partly inaccurate) derivation in particular to indicate the various assumptions of the Goldstone theorem.

The first assumption refers to the existence of a conserved current. This assumption is decisive because it expresses that there exists a symmetry of the dynamics, at all. In our proof above the existence of such a locally conserved current is necessary to conclude that the spurions are not isolated states $\delta(p^\mu)$ but can be localized to become mass zero particles. However, it is not important that $j^\mu(x)$ is really a local current. It is sufficient to require that for an arbitrarily large, but still finite volume V with the surface S, the change of the 'charge' $Q(t)$ with time within this volume is accompanied by a current $J_S(t)$ leaving through the surface, i.e.

$$\frac{d}{dt} Q_V(t) = -J_S(t). \tag{10}$$

The volume V e.g. may be a measurable region in a bubble chamber or even the volume of the bubble chamber itself. The requirement of a conserved local current $j^\mu(x)$ would mean, that this relationship holds for *any* volume, and hence also for the infinitely small volume element in which case we can write:

$$Q_V(t) = \int_V d^3x\, j^0(\mathbf{x}, t)$$

$$J_S(t) = \int_S \mathbf{ds} \cdot \mathbf{j}(\mathbf{x}, t) \tag{11}$$

If we have only the relationship (10), then upon a symmetry variation of $\phi(x)$ only within the volume V at time t', we would get

$$\frac{1}{\delta\lambda} \langle 0| \delta_{\text{sym}}^{V_1 t'} \phi(x) |0\rangle = -i \langle 0| [Q_V(t'), \phi(x)] |0\rangle = C_V(\mathbf{x}; t - t') \neq 0 \tag{12}$$

with

$$\lim_{V \to \infty} C_V(\mathbf{x}; t - t') = C = \text{const.}$$

and hence due to (10)

$$\frac{1}{\delta\lambda}\frac{d}{dt}\langle 0|\,\delta_{\text{sym}}^{V_1 t'}\phi(x)\,|0\rangle = -\mathrm{i}\,\langle 0|\,[J_S(t'),\phi(x)]\,|0\rangle = \frac{d}{dt}C_V(\mathbf{x};t-t')$$

$$= \begin{cases} 0 & \text{in region } A \\ \neq 0 & \text{outside.} \end{cases} \quad (13)$$

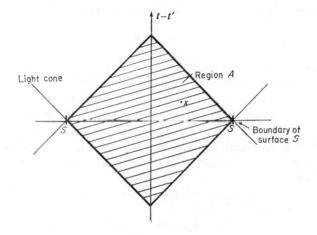

For a local theory this vanishes for x within a region A, bordered by light cones through the surface points S, i.e. for $t = t'$ if x is inside the volume. Condition (13) is sufficient to localize the spurion to a certain extent, and hence to prove the existence of a zero mass particle.

In a relativistic quantum field theory it is by no means trivial to establish the existence of a conserved current, because the construction of such currents usually involves products of field operators at the same space-time point which are rather singular objects. Consequently it is certainly not sufficient to establish the conservation laws simply on the basis of the usual (classical) substitution transformations. Nevertheless one has to keep in mind that the Goldstone theorem only breaks down if the weaker requirement (10) is violated.

An example of such an intrinsically broken symmetry is Schwinger's model of 2-dimensional quantum electrodynamics[8], where, despite of the formal γ_5-symmetry, the chirality current is not conserved. Some people call this case a locally broken symmetry[9]. However, such a definition seems only worthwhile if something is left of the symmetry at all, e.g. the weaker condition (10) which then would still imply a 'zeron'. Otherwise we should simply call this a 'no symmetry' case.

The next question refers to the existence of the generator Q. One can easily show that if the vacuum is *not* invariant under the symmetry group, i.e. if $Q|0\rangle \neq 0$, then the integral (2) will diverge. Physically speaking the 'charge' of the translational invariant vacuum state is either zero (invariant case) or infinite. In the latter case also all other states created from it by application of a finite number of field operators have also infinite charge. Hence Q itself is an infinite operator and no proper element of the Hilbert space. Hence also the unitary transformation:

$$U(\lambda) = e^{i\lambda Q} \qquad (14)$$

connected with the symmetry leads out of the Hilbert space. Vacuum states $|0\rangle_\lambda$ which are formally obtained by the transformation (14) belong to different inequivalent representations of the algebra of the field operators. They differ, so to say, by an infinite number of spurions or infrared zerons.

Although the operator Q is ill-defined, the commutator of Q with any operator localized in a certain finite region, which only was used above, is well-defined. In this case one can always work with the well-defined operator $Q_V(t)$ and go to the well-defined limit $V \to \infty$ of the commutators. It is important in this context that the commutator is assumed to vanish for large space-like separations, a property which is required for all observables in a local relativistic theory.

It was pointed out by Higgs, Englert, and Brout[10], and also by Hagen, Guralnik, and Kibble[11] that the Goldstone zerons can be avoided if there are in addition long range interactions in the theory from the outset, however, as will be seen, at a dear price. That long range forces affect the Goldstone theorem, in fact, was recognized earlier by Anderson[12] for the nonrelativistic case in particular in connexion with superconductivity where the Goldstone modes are pushed up to become the plasmons. For the relativistic case, however, it was important to recognize that long range interactions occur in connexion with gauge fields, i.e. mass zero vector fields, where the nonlocal character of the interaction becomes apparent if one uses the non manifestly covariant Coulomb gauge related to a certain space-like surface

$$\partial A - (n\partial)(nA) = 0 \qquad (15)$$

which involves the time-like vector n^μ the normal to this surface. Due to the long-range interaction (Coulomb type interaction) the commutator $\langle 0|[Q_V(t'), \phi(\mathbf{x}, t)]|0\rangle$ does *not* become time independent in the limit $V \to \infty$, since the surface current term does not decrease sufficiently fast.

At the same time the commutator is not manifestly covariant. In fact, the commutator (7) has now the more general form[10]

$$\langle 0| [j^\mu(x'), \phi(x)] |0\rangle = \frac{\partial}{\partial z_\mu} f(z) + (g^{\mu\nu}\partial^2 - \partial^\mu\partial^\nu)n_\nu g(z) \qquad (16)$$

If long range interaction are involved, then we can assume $f(z) = 0$, i.e. that *no* mass zero particle occurs, since one has instead

$$g(z) = i \frac{(n\partial)}{\partial^2 - (n\partial)^2} C \int dm^2 \rho(m^2) \Delta(z; m) \qquad \int dm^2 \rho(m^2) = 1 \qquad (17)$$

to take care of the condition (6), where the inverse operator $\partial^2 - (n\partial)^2 \to -\nabla^2$ indicates the nonlocal character of this case. In addition, the now massive Goldstone particle combines with the two massless vector bosons to form a normal 3 component massive vector field. So, no massless particles are left in the theory.

One may, of course, also use the manifestly covariant and local Gupta–Bleuler description using the Lorentz gauge, which, however, implies the introduction of an indefinite metric in the Hilbert space. In this case, the Goldstone theorem is valid and hence formally leads to a zeron, which, however, can be shown[13] to decouple completely from the physical states and eventually is eliminated, if one projects on the physical subspace of the Hilbert space. Hence also in this description there is eventually no physical zeron left.

However, it should be emphasized that the example of Higgs and others shows that a theory of this type is only causal, and hence physically acceptable, if the states created by the operator $\phi(x)$ from the vacuum can be separated completely from the physical states and suppressed. In this case one then has to check whether the symmetry, which is broken, still has a non-trivial meaning in the subspace after this projection. This does *not* seem to be the case, as we will shortly indicate in the model given by Higgs:

Higgs[10] starts out with a model originally suggested by Goldstone with the Lagrangian*

$$L_0 = \frac{1}{2}\left[\phi^{\mu*} \partial_\mu \phi + \phi^\mu \partial_\mu \phi^* - \phi^{\mu*} \phi_\mu - \frac{m_0^2}{4\eta^2}(\phi^*\phi - \eta^2)^2\right] \qquad (18)$$

for a scalar nonhermitean field operator $\phi(x)$, in which the potential has a

---

* The notation here is that essentially used by Kibble[13].

minimum for $|\phi|^2 = \eta^2 = $ const. In the groundstate hence

$$\langle 0 | \phi | 0 \rangle = \eta e^{i\alpha} \neq 0 \quad (\alpha = \text{const.}) \tag{19}$$

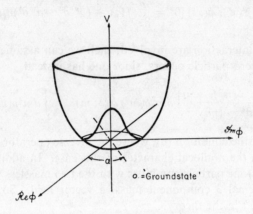

The time component of ϕ^μ are the canonical conjugate variable of $\phi(x)$. The Lagrangian is obviously invariant under the gauge transformations

$$\begin{aligned}\phi(x) &\to e^{ie\lambda} \phi(x) \\ \phi^*(x) &\to e^{-ie\lambda} \phi^*(x)\end{aligned} \tag{20}$$

which leads to a conserved current $\partial_\mu j^\mu = 0$ with

$$j^\mu = -ie[\phi^{\mu*} \phi - \phi^\mu \phi^*] \tag{21}$$

The symmetry (20) is broken by the groundstate, because

$$\frac{1}{\delta\lambda} \langle 0 | \delta\phi | 0 \rangle = ie \langle 0 | \phi(x) | 0 \rangle = ie\eta e^{i\alpha} \neq 0. \tag{22}$$

The Lagrangian may be rewritten by introducing the modular dependence $R(x)$ and the phase dependence $\theta(x)$ of the fields as new field variables:

$$\begin{aligned}\phi(x) &= R(x) e^{i\theta(x)} \\ \phi^*(x) &= R(x) e^{-i\theta(x)}\end{aligned} \tag{23}$$

with the corresponding canonically conjugate variables as the time components of the vectors $R^\mu(x)$ and $\theta^\mu(x)$. One obtains

$$\begin{aligned}L_0 &= R^\mu \partial_\mu R - \tfrac{1}{2} R^\mu R_\mu - \frac{m_0^2}{8\eta^2} [R^2 - \eta^2]^2 + \theta^\mu \partial_\mu \theta - \frac{\theta^\mu \theta_\mu}{2R^2} \\ &= L_R + L_\theta\end{aligned} \tag{24}$$

which upon variation besides the actual field equations leads to the algebraical relations

$$R_\mu = \partial_\mu R \qquad (25)$$
$$\theta_\mu = R^2 \, \partial_\mu \theta$$

which may also be inserted into the Lagrangian without harm. The symmetry transformation (20) in the new variables is now simply

$$\theta(x) \to \theta(x) + e\lambda \qquad (26)$$

with the other variables remaining unchanged, and the current (21) is

$$j^\mu = -e\theta^\mu(x) \qquad (27)$$

The groundstate condition (19) is now expressed by

$$\langle 0| R(x) |0\rangle = \eta \qquad (28)$$
$$\langle 0| \theta(x) |0\rangle = \alpha$$

The first condition (28) does *not* indicate a symmetry violation, because $R(x)$ is invariant under the symmetry transformation. The symmetry breaking condition (22) arises solely from the second condition in a rather trivial fashion:

$$\frac{1}{\delta\lambda} \langle 0| \delta\theta(x) |0\rangle = e \neq 0 \qquad (29)$$

From the latter it follows immediately by the Goldstone theorem that $\theta(x)$ is a massless field, in fact, the Goldstone zeron, which can also be directly seen from the Lagrangian, if we approximately replace $R(x) \approx \eta$. The $R(x)$ field on the other hand is connected with a particle of finite mass m_0, which can be deduced by introducing the new field operators

$$r(x) = R(x) - \eta \qquad r^\mu(x) = R^\mu(x) \qquad (30)$$

One now introduces a massless gauge field $A^\mu(x)$ by the prescription ($F^{\mu\nu}n_\nu$ is the canonically conjugate variable of A^μ):

$$L = -\tfrac{1}{2}F^{\mu\nu}(\partial_\nu A_\mu - \partial_\mu A_\nu) + \tfrac{1}{4}F^{\mu\nu}F_{\mu\nu} + L_0(\text{with } \partial_\mu\theta \to \partial_\mu\theta + eA_\mu)$$
$$= L_F + L_R + \theta^\mu(\partial_\mu\theta + eA_\mu) - \frac{\theta^\mu\theta_\mu}{2R^2} \qquad (31)$$

which obviously is invariant under the coordinate dependent gauge transformations:

$$\theta(x) \to \theta(x) + e\lambda(x) \qquad (32)$$
$$A^\mu(x) \to A^\mu(x) - \partial^\mu\lambda(x)$$

The vector θ_μ is here now obviously

$$\theta_\mu = R^2(\partial_\mu \theta + eA_\mu) = -\frac{1}{e} j_\mu(x) \qquad (33)$$

The critical commutator for the Goldstone theorem is

$$\langle 0| \, [j^\mu(x'), \theta(x)] \, |0\rangle = -e\langle 0| \, [\theta^\mu(x'), \theta(x)] \, |0\rangle \qquad (34)$$

which in the Coulomb gauge is nonlocal, and hence does not fulfil the requirements of the Goldstone theorem. In fact, one finds:

$$i\langle 0| \, [\theta_\mu(x'), \theta(x)] \, |0\rangle = \frac{(g_{\mu\nu}\partial^2 - \partial_\mu\partial_\nu)n^\nu}{\partial^2 - (n\partial)^2} \, (n\partial) \int \rho(m^2) \, dm^2 \, \Delta(z; m)$$

$$z = (x' - x) \qquad (35)$$

with $\rho(m^2) \approx \delta(m^2 - e^2\eta^2)$.

On the other hand one can also introduce the new field operator

$$B_\mu(x) = \frac{1}{eR^2} \theta_\mu = A_\mu(x) + \frac{1}{e} \partial_\mu \theta(x) \qquad (36)$$

instead of $A_\mu(x)$ (the canonically conjugate variable is still $F^{\mu\nu}n_\nu$) and obtain the Lagrangian in the form

$$L = -\tfrac{1}{2}F^{\mu\nu}(\partial_\nu B_\mu - \partial_\mu B_\nu) + \tfrac{1}{4}F^{\mu\nu}F_{\mu\nu} + \tfrac{1}{2}e^2R^2 B_\mu B^\mu + L_R \qquad (37)$$

which does not contain $\theta(x)$ any more as variable. The canonical momentum θ^μ did combine with the two transversal fields of A^μ to give a (3 component) massive vector field B^μ with mass $m_V^2 = e^2\eta^2$. If one restricts oneself to the state space produced by only applying the gauge invariant operators $F^{\mu\nu}(x)$, $B^\mu(x)$, $R(x)$ (but not $\theta(x)$!) on the vacuum, then the original symmetry no longer has meaning in this restricted state space, since all operators are (trivially) invariant under the gauge transformations. Hence this does not lead into contradiction with the general proof of Ezawa and Swieca[6].

The situation is similar, if one introduces non-Abelian groups as Kibble has shown[13]. Again the original Goldstone zerons become massive by a coupling to a corresponding gauge field by which procedure the gauge fields themselves become massive. At the same time the original symmetry transformations become meaningless in the gauge invariant observables. Only the mass zero gauge fields of the *unbroken* symmetries survive. However, their mass zero character has nothing to do with the Goldstone theorem. They are rather the leftover mass zero fields which were put in from the beginning.

To state the result more clearly, let us consider the isospin group $SU(2)$. If we break the symmetry around the x- and y-axis by the vacuum, one obtains in the simplest case of a scalar isovector field *without* gauge fields, a massive scalar boson S^0, and positively and negatively charged Goldstone zerons S^+, S^-, if for simplicity we identify the properties of the zerons with the properties of the corresponding spurions. We get other zerons by combining S^+ and S^- with positively and negatively charged spurions. If the gauge field is turned on, i.e. if one introduces a massless vector isotriplet A^+, A^0, A^-, the A^+ and A^- combines with S^+ and S^-, respectively, to give two massive vector bosons V^+, V^- and only the A^0 remains massless. Hence:

massive scalar $m = m_0$ $\qquad S^0$

massive vector $m = e\eta$ $\qquad V^+, V^-$

massless vector $m = 0$ $\qquad A^0$

The spectrum hence contains incomplete multiplets. The remaining massless gauge field is connected with the nonviolated rotation symmetry around the 3-isospin-axis. If *all* symmetries are broken, there will be *no* massless field left.

If one applies this procedure to a $SU(3)$ invariant theory and subsequently breaks this symmetry by strong, electromagnetic and weak interactions, the only remaining mass zero particle is the photon[14]. Otherwise, however, the model has very unrealistic features, in particular again the original high symmetry has completely lost its meaning.

Perhaps one has to be somewhat more careful with the statement that these symmetries are physically meaningless because all observables are left invariant. Also the electric and baryon number gauge transformations are such symmetry transformations which leave the observables invariant, but nevertheless have important physical consequences. In fact, this invariance leads to the superselection rules. However, it appears, that the strict validity of these transformations is the decisive difference. A *broken* symmetry on the other hand can only be sensed, if the theory contains *some* observables which are *not* invariant.

One may believe that at least in the case of a non-Abelian symmetry group like $SU(2)$, there are some objects left in the theory (namely the incomplete multiplets), which do not have a trivial transformation character. According to the construction this, however, is not the case. For example, neither the V^+, V^- nor the S^0, nor the A^0 transform like an isotriplet any longer.

This is, in fact, a general deficiency of all multiplets which are broken by an asymmetrical vacuum, and is not characteristic of the Higgs case. It was Umezawa who particularly stressed this point[15]. Let us again consider the broken $SU(2)$ isospin symmetry case without the gauge fields. The vacuum in this case could be hypothetically imagined as a large isoferromagnet with nonvanishing magnetization in the z-direction. The energy levels of a particle with isospin $\frac{1}{2}$ will split up consequently into two nondegenerate mass components, depending on the isospin orientation. This, however, is not a correct description. If we imagine, for the moment, that the vacuum is very large but finite, we may characterize it by a large isospin I_v. The non-degenerate mass doublet is then described by states of total isospin $I = I_v \pm \frac{1}{2}$. i.e. as energy levels corresponding to *different* irreducible representations of the symmetry group. If one performs an isospin rotation the levels hence do *not* transform into each other, because they differ in energy, but rather transform into themselves plus spurion contributions. The spurions correspond to the infrared limit of the Goldstone zerons. The whole reaction to our symmetry transformation seems to consist in a rearrangement of the isospin in the groundstate. In particular, if we separate from the field operator a part which, like $R(x)$ described above, does not participate in the broken symmetry transformations and connect the physical particle with it, then the physical particle will behave like an isosinglet under the I_1 and I_2 rotations, i.e. its isospin degree of freedom will appear to be 'frozen'. The whole transformations only add terms to the rest fields, which, as Umezawa remarks, merely change the Bose–Einstein condensation of the Goldstone zerons. In this language all the components of a nondegenerate multiplet behave like singlets under the relevant transformations. The variant part, a spurion part, is disconnected and combines with the BE-condensation part of the Goldstone bosons.

This indicates that the original symmetry transformation no longer connects the components of a split multiplet. There may, however, exist *another*, although weakly time-dependent transformation, which *does* connect the components of a multiplet in the expected way. In our example, it will consist of the isospin transformation which rotates the isospin only of the particle and not the large isospin of the vacuum. Whether this new transformation is a sufficiently good approximate symmetry transformation, i.e. is sufficiently time-independent, will depend on the strength of the coupling of the isospin of the particle to the vacuum isospins relative to their mutual coupling. If the particle coupling is strong a description as an isosinglet or an isotriplet may be more appropriate. Biritz[16] has given an example of such anomalous multiplets in a model of a ferromagnetic chain. In connexion with the nonlinear spinor theory strange particles were interpreted as such anomalous isospin multiplets[17].

In the non-Abelian models of Higgs and Kibble the originally degenerate multiplets after breaking of the symmetry become nondegenerate and even partially incomplete. It is hard to see how in this case an approximately valid symmetry transformation can be found which would transform these objects as members of the original irreducible representation. Obviously, a theory which approximately retains *only* the multiplets, but not the corresponding Clebsch–Gordan coefficients and selection rules would be completely useless.

There are still a number of other questions which should be studied in detail. One question is, what happens to the description of the isospin rotations around the x- and y-axis in the state space, if the superselection rules for the electric charge are established, because then these transformations, even if performed locally, would connect different invariant sectors. An investigation of this question may perhaps reveal that Higgs' suggestion may be physically applicable, after all, in the case where a gauge symmetry is left intact, because the leftover massless gauge field would enforce a superselection rule.

The isospin symmetry in elementary particle physics has been conjectured long ago by Heisenberg and coworkers[2] to be a possible candidate for an exact dynamical symmetry which is broken by the groundstate, because its violation is accompanied by and phenomenologically attributed to the mass zero photon. Unfortunately, however, the Goldstone 'zerons' in connexion with violation of the $SU(2)$ isospin group are, as we have seen, scalar, and, in the usual description, charged objects. Intuitively they are the magnons, the Bloch spin waves, of an infinitely large isoferromagnet at zero temperature. On the other hand it is quite clear that the isoferromagnet would not be an adequate description for the physical situation, because such a vacuum state would also violate CPT-invariance. One can easily see this if we imagine e.g. a π-meson in such an isoferromagnet, which would naturally split up into a mass triplet and hence violate the requirement that π^+ and π^- are antiparticles. This indicates that in this case, one has to employ a more complicated way to break the $SU(2)$ symmetry.

A more appropriate model for the isospin violating vacuum state in elementary particle physics would be a model in which at every lattice point there is a particle and an antiparticle with their isospins pointing essentially in opposite directions. There would be no resultant polarization (charge), but the polarization of the particle and antiparticle subsystems would be very large and distinguish a certain direction. Under CPT such a system would transform into itself. A state of this type has formally some similarity with the groundstate of an antiferromagnet, where the spins of neighbouring particles try to be antiparallel. One obtains such a groundstate, if one introduces forces which tend to align the spins of like particles and antialign spins of unlike particles. The groundstate of such a system is a very complicated object. Biritz and Yamazaki[18] have recently started to investigate this model in an approximation where the antialigning forces between the particles and antiparticles are considered weak in comparison with the aligning forces between like particles. In lowest approximation one obtains essentially a double isoferromagnet, a particle- and antiparticle-isoferromagnet oriented in opposite directions. This describes a particular superposition of states with isospin $I = 0, 1 \cdots$ with zero charge, i.e. without net polarization in the 3 direction. With the particle–antiparticle forces fully switched on, the situation gets very complicated, because these forces tend to form local singlets which upset the double ferromagnetic ordering by flipping a certain number of particle-antiparticle pairs. The exact groundstate has not been worked out as yet. The model has to be studied in three dimensions, since, similar to the antiferromagnet[19], the zero-point fluctuations do not permit a long range ordering in one and two dimensions.

In this iso-antiferromagnetic model there seems to appear a new uncharged Goldstone mode which is connected with a localized flipping of a particle–antiparticle pair. There is a way to write this mode as a projection operator

$$|k\rangle \sim \sum_n \tfrac{1}{2}(1 + \tau_3) e^{ikn} |0\rangle$$

which senses localized flipped particle–antiparticle pairs and hence has some similarity with the photon. An interesting question, however, is whether one can avoid now the charged Goldstone modes which are suggested still to show up in the usual interpretation on the basis of a general proof of the Goldstone theorem in the algebraic approach.

One actually would suspect that such charged modes, at least in the usual interpretation, should arise in connexion with a rotation of the particle–antiparticle lattice as a whole. However, one realizes that a rotation of the lattice, which has vanishing magnetization, after the rotation leads to a state for which the expectation value of the magnetization in the z-direction is still zero. Hence one may expect that the charged modes do not occur in the same way as in the ferromagnet.

Of course, there would still remain the question how the Goldstone zeron can be endowed with spin without breaking the Lorentz group by the vacuum, which actually would be required in this case by the Goldstone theorem[20]. The hope is here, that actually the Coulomb force, which is scalar, is inferred by the Goldstone argument, and that the photons only follow indirectly from it by locality and Lorentz invariance. This all has still to be investigated.

The question whether the neutrino may follow from some kind of a Goldstone argument must be completely denied at present. This becomes particularly clear in Umezawa's argumentations, where the Bose–Einstein condensation of the zerons appear to be crucial.

Before closing I wish to make a few remarks about the possibility to directly observe the underlying dynamical symmetries. Up to now we have argued that the existence of a local conservation law can only be indirectly inferred through the appearance of the Goldstone zerons, at least, if no long range forces are present from the beginning. The question arises whether there is not a direct way to establish the local conservation law. After all, it states that for every given volume the time-change of the 'charge', e.g. the first or second component of isospin, must be connected with a corresponding current going through the surface of this volume. If we find in a bubble chamber experiment that in a certain process isospin is violated, then—in our interpretation—this can only mean that our book-keeping is incorrect, that some isospin must have leaked out of the chamber unaccounted. There are, in fact, two reasons why our conventional book-keeping could be wrong:

(1) mass zero particles carrying isospin with an energy smaller than our energy resolution may have escaped our observation (infrared problem);

(2) The interacting particles were erroneously assumed to be exact eigenstates of isospin. Already their mass splitting, however, indicates that this can be only approximately true. So e.g. the ω^0 has small admixtures of the quantum numbers of ρ^0, the π^0 those of η, etc.

I finally want to remark on the non-leptonic weak interactions which phenomenologically can be successfully formulated as isospin $\frac{1}{2}$-spurion emission processes. In a theory with an isospin degenerate vacuum this formal description may even have a realistic foundation, because, as Umezawa has indicated, such a spurion is connected with the Bose–Einstein condensation of the Goldstone boson. It indicates a transfer of intrinsic quantum numbers to the vacuum as a whole. Such a description seems to have some formal similarity with the Mössbauer effect, where apparently the recoil momentum of the γ-emitting nucleus is transferred to the crystal as a whole, and hence seems to be locally lost. Weisskopf[21] has demonstrated, due to the high zero-point fluctuations connected with a local measurement, that there is no observable violation of the causality principle.

In conclusion, I wish to stress that, to our knowledge, there exists no case in which a dynamically valid symmetry in a relativistic, causal theory can be broken by the groundstate in an observable way without involving mass zero particles. The main problem in our opinion in dealing with the Goldstone theorem in relativistic theories seems to be connected with the physical interpretation of the assumptions which go into it—in particular, what we mean by a symmetry and its violation—and the physical interpretation of the consequences we derive from it—in particular whether we retain an approximate symmetry for the non-degenerate one-particle states which originally belonged to a single multiplet of the symmetry group and what are the symmetry properties of the zerons. Probably all these questions can only be decided by actually carrying out dynamical calculations.

References

(1) Baker, M., and Glashow, S. L., *Phys. Rev.*, **128**, 2462 (1962).
(2) Heisenberg, W., *Rev. Mod. Phys.*, **29**, 269 (1957). Heisenberg, W., *Proc. Ann. Intern. Conf. High Energy Phys. CERN*, 119 (1958). Dürr, H. P., Heisenberg, W., Mitter, H., Schlieder, S., and Yamazaki K., *Z. Naturforsch.*, **14a**, 441 (1959). See also Heisenberg, W., *An Introduction to the Unified Theory of Elementary Particles*, Wiley, London, 1967; German ed. Hirzel Verlag, Stuttgart, 1967.
(3) *See* e.g. Wagner, H., *Z. Physik.* **195**, 273 (1966).
(4) Goldstone, J., *Nuovo Cimento*, **19**, 154 (1961): Goldstone, J., Salam, A., and Weinberg, S. *Phys. Rev.* **127**, 965 (1962).
(5) Bludman S. A., and Klein, A., *Phys. Rev.*, **131**, 2364 (1963).
(6) Kastler, D., Robinson, D., and Swieca, A., *Commun. Math. Phys.*, **2**, 108 (1966); *see also* Ezawa, H., and Swieca, J. A., *DESY preprint*, **66/38** (1966).
(7) Bludman, S. A., *Proc. of seminar on unified theories of elementary particles, Max-Planck-Inst. für Physik und Astrophysik, Feldafing p.* 36 (1965).
(8) Schwinger, J., *Phys. Rev.* **128**, 2425 (1962); Theoret. phys. IAEA Wien 1963, p. 89.
(9) Maris, Th. A. J., *et al.*, preprint 1967.
(10) Higgs, P. W., *Physics Letters* **12**, 132 (1964); Higgs, P. W., *Phys. Rev.* **145**, 1156 (1966); Englert, F., and Brout, R., *Phys. Rev. Letters* **13**, 321 (1964).
(11) Guralnik, G. S., Hagen, C. R., and Kibble T. W. B., *Phys. Rev. Lett.* **13**, 585 (1964); see also Kibble, T. W. B., Proc. of the Intern. Conf. on Elementary Particles (1965) p. 19; and Proc. of the Intern. Theor. Physics Conference on Particles and Fields (1967).
(12) Anderson, P. W., *Phys. Rev.* **130**, 439 (1963).
(13) Kibble, T. W. B., Imperial College Preprint ICTP/66/25.
(14) Higgs, P. W., *Phys. Rev. Letters*, **13**, 508 (1964).
(15) Umezawa, H., *Nuovo Cim.* **40**, 450 (1965); Leplace, L., Sen, R. N., Umezawa, H., *Nuovo Cim.* **49**, 1 (1961); Sen, R. N., Umezawa, H., *Nuovo Cim.* **50**, 53 (1967).
(16) Biritz, H., *Nuovo Cim.* **47**, 581 (1966).

(17) Dürr, H. P., and Heisenberg, W., *Nuovo Cim.* **37,** 1446 (1965); Dürr, H. P., and Heisenberg, W., *Nuovo Cim.* **37,** 1487 (1965);
(18) Biritz, H., and Yamazaki, K., to be published.
(19) Merwin, N. D., Wagner, H., *Phys. Rev. Letters* **17,** 1133 (1966).
(20) Bjorken, J. D., *Annals of Physics* (N.Y.) **24,** 174 (1963).
(21) Weisskopf, V. F., *Boulder Lectures in Theoretical Physics*, Vol. III, p. 79 (1960).

Discussion on the report of H. P. Dürr

S. Weinberg. I would like to emphasize a little more strongly the applications of the Goldstone theorem to the real world. First, let me mention in passing, that there is one other classical example of a Goldstone mode besides the ones that are always quoted, i.e. the hose instability of a charged particle beam in a uniform plasma. Coming back to the strong interactions, I would say that the greatest triumph of the Goldstone theorem is that it gives a 'raison d'être' for the pion as an almost massless particle. From this point of view, it is not important whether the Goldstone theorem has been rigorously proved; the important thing is that it tells us how the strong interactions could keep the pion mass so small.

I was also very impressed with the suggestions of Higgs, Kibble, and others, and would like to point out that the ρ and A_1 mesons afford a good example of these ideas. Of course isotopic spin is not broken, so the ρ meson mass has to be put in at the beginning, and then chiral symmetry 'breaking' gives an additional mass to the A_1, which in fact agrees with experiment. (Because the ρ has a bare mass, the Goldstone bosons don't go away; they are just the pions.) We can also ask whether these ideas can be applied to unify the electromagnetic and the weak interactions. If we restrict ourselves to the observed electronic leptons, then there is just one way that this can be done: there must be a massless photon, a massive charged intermediate meson, and a heavier intermediate neutral meson. The neutral intermediate boson has observable effects, notably that in electron–neutrino scattering the axial-vector $(\bar{e}e)(\nu\bar{\nu})$ coupling is $\frac{3}{2}$ what we would expect from the old calculation of Feynman and Gell-Mann.

This raises a question that I can't answer: Are such models renormalizable? You start with a Yang–Mills type Lagrangian which is renormalizable, and re-order the perturbation theory by redefining the fields. I hope someone will be able to find out whether or not the resulting Lagrangian is a renormalizable theory of weak and electromagnetic interactions.

F. Englert. With regard to the renormalizability of gauge vector mesons in the presence of broken symmetry, Brout and I (*Phys. Rev. Letters* **13**, 9 (1964) and *Nuovo Cim.* **43**, 244 (1966)) showed that the propagator of the massy vector mesons is $[g_{\mu\nu} - q_\mu q_\nu/q^2](q^2 - \mu^2)^{-1}$, μ being the induced mass of the gauge field. The term in $(q_\mu q_\nu/q^2)$ having a singularity at $q^2 = 0$ is due to the Goldstone boson contribution to the vector meson propagator. The term enters in this way as a consequence of the Ward identity and is not therefore an effect due to an approximation. Therefore the answer to Weinberg's question is that these massy vector gauge fields constitute a renormalizable field theory.

H. P. Dürr. I wish to apologize that I haven't mentioned, at all, the theories where chirality is broken, which was first treated by Nambu. Of course, it appears that chirality is fundamentally broken by a slight amount such that the π-meson acquires a small mass. One can nevertheless, as Weinberg has pointed out, actually recover many relationships of the Goldstone situation.

F. E. Low. Does a Goldstone boson lie on a trajectory?

H. P. Dürr. I don't know, but I would suspect that the answer may depend on the particular dynamics one has to deal with.

R. Brout. At least in the Nambu model where the π-meson is an $(N\bar{N})$ bound state, it would appear unlikely that the π-meson reggeizes. When one turns on small bare mass, the matrix elements of $\partial_\mu j_{\mu 5}$ do not tend to zero at ∞ momentum transfer, but rather to the matrix elements of $m_0 \bar{\psi} \gamma_5 \psi$, m_0 being the bare mass. Thus there is no unsubtracted dispersion relation for the pseudoscalar form factor, as would be expected from reggeization of the pion and a no subtraction hypothesis. In the case that $m_0 = 0$ and $\partial_\mu j_{\mu 5} = 0$, the pion is the Goldstone pole and remains the pseudoscalar pole in the form factor, no matter how high the momentum transfer. This follows from the Ward identity. Again the behaviour is contrary to Regge behaviour.

R. E. Marshak. Since we are working in a rather speculative domain, I should like to make a remark which illustrates the different approaches to chirality symmetry taken by Weinberg and by some of us who have worked exclusively with fermion fields. I find it extremely strange and suggestive that we not only have three lepton fields in close analogy to the three quarks which seem to structure the hadrons, but we also have 3 symmetry breakings from the $SU(3) \otimes SU(3)$ level, down to $SU(3)$, down to $SU(2)$ and finally down to $SU(1) \times Y$. In a sense, the muon and electron are behaving like the quarks which produce the hypercharge and electromagnetic splittings respectively among hadrons. In a more serious vein, Weinberg is willing to tolerate a non zero mass pion as a sort of pseudo-Goldstone particle. One might have a chance to explain the 3 broken symmetries in terms essentially of lepton currents compounded out of the 3 objects we know with different masses: zero mass ν, electromagnetic mass e, and muon mass which is of the order of (hypercharge) $SU(3)$ breaking. Perhaps this conjecture can already be excluded by what is known about simultaneous symmetry breakings and the generalized Goldstone theorem.

H. P. Dürr. At present I cannot see how a connexion can be made between the Goldstone theorem and the leptons. Perhaps there might be a way to establish a connexion between weak currents and the breaking of symmetries if one uses some kind of a mechanism as suggested by Higgs and others which make use of long range forces in the original Lagrangian.

W. Heisenberg. In connexion with the Goldstone theorem I want to stress two points. The one concerns the interaction of a particle with the groundstate. If the breaking of the symmetry is seen experimentally in the splitting of mass levels in a multiplet, then in the Goldstone case, the splitting must be due to this interaction of the particle with the groundstate and its collective modes. Therefore if one were to try to explain the violation of $SU(3)$ by a Lagrangian exactly invariant under $SU(3)$ and an asymmetrical groundstate, there should not only exist the collective modes of mass zero in the groundstate, but these modes should also interact strongly with the particles—against existing experimental evidence. With respect to $SU(2)$ the situation is much better. There we have the Coulomb field and the photon, and their interaction is just of the correct order of magnitude for explaining the mass splitting in the iso-multiplets.

The second point concerns the projection operator in the definition of a Goldstone particle. These particles may be considered as localized spurions. Now a spurion, being the change from one vacuum to another vacuum, means some change which is—with equal probability—spread out over the whole space. If one then constructs a projection operator which picks out just those points in space where the change has occurred, then by means of this operator one can localize the spurion, i.e. construct a Goldstone particle. Therefore the occurrence of a projection operator like e.g. $\frac{1}{2}(Y + \tau_3)$ in the Gell-Mann–Nishijima rule is a strong indication for the Goldstone-character of the corresponding particle or field.

R. E. Marshak. How can you hope to demonstrate that the photon is the Goldstone particle in $SU(2)$ symmetry breaking when it has the wrong quantum numbers?

H. P. Dürr. There may be still some hope to change the internal quantum numbers of the Goldstone boson by coupling on spurions, as Heisenberg has suggested. The Lorentz properties would be all right if one could show that the Goldstone object is really the scalar Coulomb force from which then the photons would arise by a secondary step from Lorentz invariance.

Personally, I certainly would take the suggestion of Higgs and others as an attractive possibility, provided one can find some way out in the difficulties which seem to occur there.

G. Källén. If it is true that the π-meson is a Goldstone particle with approximately zero mass then I would like to ask: what is the situation for the other mesons usually classified to be in the same octet?

Do they all correspond to Goldstone particles, but with some of the masses more equal to zero than others? Alternatively, one could declare that the octet classification is an accident and that the π-meson is basically different from the other pseudoscalar mesons. If the first alternative is preferred by the official point of view my question is: Do all the pseudoscalar mesons correspond to the same broken symmetry or are they related to different broken symmetries? If so, which broken symmetry corresponds to which particle?

S. Weinberg. I don't know what the official view is, but the consensus seems to be that it is better not to think about the strange particles if you want to go on thinking you understand what is going on. Nevertheless, Glashow and I have looked at what happens to Goldstone's theorem if you include $SU(3) \times SU(3)$ symmetry breaking with a specific assumption as to how the symmetry breaking term transforms. The result is very weak, i.e. one inequality among the masses of the would-be Goldstone bosons (the pseudoscalar nonet and the unobserved kappa meson), which may be true. In addition, it should be noted that the calculations of KN and $\bar{K}N$ scattering lengths which use the idea of a partially conserved strangeness-changing current are in pretty good agreement with experiment.

MATHEMATICAL ASPECTS OF QUANTUM FIELD THEORY

R. Haag

Institute for Theoretical Physics, University of Hamburg,
Hamburg, Germany.

If one took an opinion poll among the participants of this conference as to what is 'Quantum Field Theory' and which mathematical disciplines are most relevant to it, I am sure one would obtain a wide spectrum of opinions. It would clearly be impossible to discuss all these aspects within one hour. Therefore, I shall not attempt any systematic treatment of the topic but single out a few questions which during the past years have seen a fruitful collaboration between mathematicians and physicists. Indeed the possibility of such a collaboration today is in itself quite remarkable. The physicist tends to regard mathematics as a stockpile of tools. He knows the older ones and is mostly contemptuous of the newer ones because they were not constructed to solve his problems. The mathematician has a different view of his role. He will be interested in a problem of physics only if he understands the conceptual background and finds in it a natural challenge for the development of mathematical ideas or if he finds examples for structural relations he has studied before in an abstract context. I learned this lesson 10 years ago when I went with a detailed list of questions to see a mathematical colleague and received the advice to go and look for a 'tame mathematician'.

The mathematical background of any reasonably complete theory of elementary particles will of course be functional analysis. The quantity to be determined, the 'unknown', is a functional i.e. an infinite set of functions or, alternatively, a function of infinitely many variables. This is quite evident because even in the most economical description of elementary particle processes, we want to know the infinitely many amplitudes for all possible scattering and production processes. Each such amplitude is a function of several momentum variables. All of them are coupled together (e.g. by unitarity relations). Similarly, the solution of a quantum field theory can be described by an infinite system of functions. In the fifties several choices for such a hierarchy of functions have been proposed and studied; the best known examples are the Feynman amplitudes, the Wightman functions and the retarded functions. In each case one finds a number of properties which the individual functions of the system have to possess due to general physical principles (Lorentz invariance, locality, positive metric and positive energy). The study of these constraints on the individual functions has sometimes been called the 'linear programme'. It was originally believed to be some reasonably simple preliminary exercise before one could tackle the tough problem of the interrelations between the different functions (of one hierarchy) which result in part again from the general principles and in part from detailed dynamical equations. It turned out, however, that already the linear programme is a formidable task for all but the lowest two functions in a hierarchy. Probably most colleagues will agree now that this strategy is impracticable without a deeper understanding of the general structure of the theory and without additional information about the quantitative aspects. The main point I want to stress here is that—irrespective of the point of view we have about the ultimate frame in which a theory of elementary particles should be cast—we are dealing with an area of mathematics in which there are very few computational techniques known. Functional integration is developed only for an extremely restricted class of functionals; functional differential equations to which a solution is obtainable in analytic form are extremely rare and probably uninteresting for our purposes. Approximation methods with a moderate amount of success have all relied on the fortuitous circumstance that in some cases there is a representation of the functional by a hierarchy of functions such that all but the first few functions can be neglected. Examples are

(i) Variational method in some cases of the nonrelativistic many body problem (infinitely extended medium with finite mean density). Using the truncated Wightman functions (uncorrelated parts of the Wightman functions) as the hierarchy, all functions depending on more than two points are neglected. The lowest functions are determined then by minimizing the expectation value of the Hamiltonian. This leads to an (approximate) description of the medium in terms of a collection of uncoupled normal modes, corresponding to elementary excitations (free quasi-particles).

(ii) Perturbation methods.

(iii) Tamm–Dancoff method and Bethe-Salpeter equation.

The combination of such methods with empirical information produced some definite progress both in the understanding of elementary particle physics and in the nonrelativistic many body problem. The time scale for such progress is however rather long, if compared to the revolutionary development of physics during the first 30 years.

My introductory remarks were made mainly for the purpose of suggesting the reason for this slow speed. It is primarily due to the complexity of a problem involving infinitely many degrees of freedom and not to a scarcity of bold and ingenious ideas. Even in the nonrelativistic many body problem, where there is no need for any new ideas concerning the fundamental dynamic equations or the mathematical frame because we believe we have had an adequate formulation of the problem for almost 40 years, we are only beginning to be able to relate some of the most striking experimental phenomena to these basic equations.

After these negative remarks let me come to some positive aspects and to the main subject of my talk. There is, of course, a large body of mathematical knowledge in functional analysis which is not concerned with numerical computational techniques, but with the classification of structures. Such results also have relevance to physics. They help us to decide whether an assumed theoretical framework is capable of describing gross qualitative features of experience. I want to give some examples for this.

The theoretical framework which will be assumed is that of a 'local quantum theory of an infinite system'. It can be specialized to quantum field theory by adding the requirements of Lorentz invariance and relativistic causality. It also underlies the non relativistic many body problem. Therefore, it gives a good basis for the very interesting comparison between phenomena in elementary particle physics with those in many body systems. No further apology for the assumption of this framework will be offered. The essential assumptions are the following:

(1) We assume that the basic mathematical quantities of the theory are elements of an algebra (denoted by $\mathscr{R}$). This means that whenever A and B belong to $\mathscr{R}$ then the following operations can be performed and give other elements of $\mathscr{R}$:

> linear combination with complex coefficients $\alpha A + \beta B$
> product AB
> adjoint A^*

The product shall be associative but not commutative. The usual laws which connect addition, multiplication and adjoint (familiar from matrix algebras) shall hold. Since $\mathscr{R}$ will have an infinity of linearly independent elements it is necessary to endow it with some topology. The simplest possibility (and one which at least in the case of the many body problem is natural and useful) is to use a norm topology. Thus each element A will have a norm $\|A\|$ and the algebra shall be complete with respect to this norm topology (Cauchy sequences in $\mathscr{R}$ converge towards elements of $\mathscr{R}$).

It is then, from the physical point of view, essentially no restriction to assume further that the norm satisfies the condition

$$\|A^*A\| = \|A\|^2$$

which is a familiar property of the norms of operators in a Hilbert space. With this condition $\mathscr{R}$ becomes an (abstract) C^*-algebra.

Physically, the selfadjoint elements of $\mathscr{R}$ (or at least a sufficiently large part of them) will be interpreted as '*observables*'. Therefore $\mathscr{R}$ will be called the algebra of observables. A physical '*state*' can be characterized by the collection of expectation values of the observables. Any numerical function assigning a complex number $\omega(A)$ to the algebraic element A will be called a state if it satisfies the two conditions

(a) linearity $\quad \omega(\alpha A + \beta B) = \alpha\omega(A) + \beta\omega(B)$
(b) positivity $\quad \omega(A^*A) \geq 0 \quad$ for all A.

These conditions ensure that with the usual probability interpretation all probabilities come out positive. The assumptions described thus far are one way of expressing the general mathematical substratum of Quantum Physics. It is, of course, not claimed that this structure is sacred. But so far no indication of its failure has been seen nor has any natural and reasonably complete alternative scheme for a theory of measurement been proposed.

(2) Next we assume a relation of the algebraic structure to space-time. Specifically, for any finite space-time region o there shall be a subalgebra $\mathscr{R}(o)$, interpreted as the algebra generated by those observables which can be measured within the region o. The total algebra $\mathscr{R}$ shall then be the norm completion of the union of all the local subalgebras:

$$\mathscr{R} = \bigcup \overline{\mathscr{R}(o)} \tag{1}$$

Statement

A physical theory, including interpretation, is given once we fix an assignment of a subalgebra $\mathscr{R}(o)$ to every space-time region o. In other words, the correspondence

$$o \rightarrow \mathscr{R}(o) \tag{2}$$

fixes the theory.

Explanation

First, some important corrections to the statement must be made. The total algebra $\mathscr{R}$ resulting from (2) and (1) must be 'simple', i.e. it should not contain any 2-sided ideals. Otherwise, the theory is not completely specified. It becomes specified then after equating the elements of a maximal 2-sided ideal $\mathscr{T}$ to zero i.e. by taking instead of $\mathscr{R}$ the quotient algebra $\mathscr{R}/\mathscr{T}$ as the algebra of observables. Apart from this qualification, the statement rests on two remarks. On the one hand one expects that the algebra $\mathscr{R}$ will allow many unitary inequivalent irreducible representations by operators in Hilbert spaces. But one can show that a distinction by unitary inequivalence is too fine to be measurable and that all possible Hilbert space representations of a simple algebra are physically equivalent. The other remark is that it is unnecessary to know a priori how different mathematical elements A_i which are based on the same space-time region are realized by the builders of hardware. On the side of the experimentalist, the development of an efficient and selective detector is a lengthy process involving much trial and error. The question as to 'what' this instrument measures is ultimately only answerable in terms of the results of measurements for varying geometric configurations of a collection of such instruments. The physical interpretation of a theory in which no further information is furnished than the space–time regions on which the various observables are based follows exactly the pattern of the experimental procedure sketched above.

Given a correspondence (2) which leads by (1) to a simple algebra we have a theory. Of course, such a model theory will describe a world which is qualitatively completely different from ours unless the correspondence (2) satisfies a number of further requirements. Such properties are:

(3) Symmetries

Within this context a symmetry of the theory will be understood as an automorphism of $\mathscr{R}$ (a mapping $A \to A_1$, conserving all algebraic relations) with the additional restriction that the image of any local subalgebra (corresponding to a finite region o), will again be a local subalgebra $\mathscr{R}(o_1)$. The mapping $o \to o_1$ associates with each symmetry a point transformation of space-time. It is clear that this associated group of point transformations in space-time must be the underlying geometrical symmetry group i.e. the Poincaré group (translations plus Lorentz transformations) or, alternatively, the inhomogeneous Galilei group. The subgroup of symmetries which leaves each o unchanged is called the internal symmetry group. The total symmetry group is an extension of the geometric group by the internal symmetry group.

(4) Locality and causality

For completeness sake I shall write down the two causality assumptions which are usually made:

(a) If we consider only regions whose points have time coordinates between t and $t + \Delta$ and denote by $\mathscr{R}_{t,\Delta}$ the algebra generated in analogy to (1) by these special regions, then

$$\mathscr{R}_{t,\Delta} = \mathscr{R} \quad \text{for arbitrary } t, \Delta.$$

(b) In the relativistic case algebras of regions which lie space-like to each other shall commute. In the nonrelativistic case with short range forces a qualitatively similar assumption can be made.

I shall not make any direct use of these two assumptions in the remainder of my talk but instead consider some related properties of the algebra which lead to a few simple and interesting consequences for the states. In this connection I want to quote a few mathematical theorems and therefore I should be careful to be more precise than usual with my definitions. For simplicity I shall however be more restrictive than necessary.

Consider a C^*-algebra $\mathscr{R}$ and a finite parametric Lie group of automorphisms $\mathscr{G}$ acting on it. $\mathscr{G}$ shall be non compact so that $g \to \infty$ is meaningful (g denotes a group element). In applications $\mathscr{G}$ will be usually the 3-parametric translation group in space or the one parametric translation group in time. We say that the system $\mathscr{R}, \mathscr{G}$ is '*asymptotically Abelian*' if for every pair of elements A, B from $\mathscr{R}$ the commutator

$$[A, B_g] \to 0 \quad \text{as } g \to \infty$$

B_g is, of course, the image of B under the automorphism g. Secondly, a state ω is called invariant under g if

$$\omega(A_g) = \omega(A).$$

Properties of asymptotically Abelian systems have been studied in a series of papers in the past two years. See ref. (1), (2), (3), (4), (5). I quote some theorems:

Theorem I

Let $\mathscr{R}, \mathscr{G}$ be asymptotically Abelian. Then the set of invariant states under $\mathscr{G}$ is a simplex.

What does that mean and why is it interesting? In general the set of (normalized) states forms a convex body as indicated in the 2-dimensional picture below. Taking any two states ω_1 and ω_2, the straight segment between the two points represents all the states which can be obtained as mixtures between ω_1 and ω_2:

$$\omega = \lambda\omega_1 + (1 - \lambda)\omega_2; \quad 0 < \lambda < 1.$$

Therefore the points on the curved part of the boundary and the corner points between straight boundary segments correspond to pure states, the others to mixtures. In our example the dimensionality of normalized states is 2. Omitting the normalization condition, we have a 3-dimensional state space corresponding to a cone of which I have drawn a cross section. There are, however, infinitely many pure states. It is immediately seen that as soon as the number of different pure states exceeds the dimension of state space, the decomposition of an arbitrary state into its pure components is not unique. This is the usual situation in quantum mechanics and this fact is closely related to the impossibility of attributing an objective meaning to the 'state of an individual system' (a meaning which does not depend on whether an observation has been made or whether the

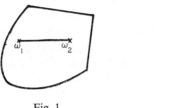

Fig. 1

Fig. 2

system is regarded as a member of statistical ensemble). A simplex on the other hand is drawn for our example in Figure 2. The number of pure states equals the dimensionality of the state space. Hence there is a unique decomposition of an arbitrary state into its pure components. We have the situation of classical physics; the pure states may be regarded as mutually exclusive possibilities. One does not get into logical contradiction if one claims that irrespective of our observations, an individual system must always be in one of these pure states and a general ensemble is described by a (classical) probability distribution over the pure states. Theorem I tells us that in an asymptotically Abelian system, the subset of *invariant* states has this classical structure. Therefore, there is a unique decomposition of an arbitrary invariant state into 'extremal invariant states'. These extremal invariant states or, synonymously 'ergodic states' are in general not pure states of the algebra $\mathscr{R}$ but they can only be decomposed into noninvariant components. Ergodic states are pure with respect to the algebra generated by $\mathscr{R}$ and $\mathscr{G}$ together. The theorem tells us that there is a superselection rule between any pair of ergodic states.

It is reasonable to expect that the algebra of observables is asymptotically Abelian with respect to time translations. Then, ergodic states with respect to the time translation group are natural candidates for thermodynamic equilibrium states (including the special cases of zero temperature and the vacuum state in field theory). Among the many general properties of ergodic states let me mention only one:

Theorem II

In an ergodic state the average value of long range correlations is zero, i.e.

$$\lim_{G \to \infty} \frac{1}{\mu(G)} \int_G (\omega(A_g B) - \omega(A)\omega(B)) \, d\mu(g) = 0$$

Here $d\mu(g)$ is the natural (Haar) measure on the group.

I mentioned these theorems as an example demonstrating the impact of simple structure properties of the algebra on qualitative features of the manifold of states. A finer analysis of the relation between algebraic structure and properties of states leads to many questions for which only partial answers are known. They involve the problem of stability (existence of a ground state, second law of thermodynamics for local perturbation from ergodic states); existence of stable single particle states; possibility of a complete particle interpretation of the states resulting from local perturbation of the vacuum; local superselection rules (gauge invariances). Let me make a few comments about the last questions.

Consider a pure state ω. We call another pure state ω' coherent with ω if there is an element B in the algebra such that

$$\omega'(A) = \omega(B^*AB).$$

If two pure states are not related in this way, then there is a superselection rule between them. A maximal collection of mutually coherent pure states will be called a sector. One reason for the appearance of a multitude of different sectors was mentioned before in connection with ergodic states. Two states are certainly in different sectors if they differ asymptotically at large spatial distances, i.e. if

$$\lim_{x \to \infty} (\omega'(A_x) - \omega(A_x)) \neq 0$$

Let us therefore consider only the subset $\mathscr{F}$ of those pure states which coincide asymptotically with the vacuum state ω_0:

$$\omega \in \mathscr{F} \quad \text{if} \quad \lim_{x \to \infty} \omega(A_x) = \omega_0(A).$$

For the algebra of a neutral free field $\mathscr{F}$ consists of a single sector. For the algebra generated by the current operators of a free Dirac field $\mathscr{F}$ has a denumerable set of different sectors. They are distinguished by the charge quantum number. In general, we may expect $\mathscr{F}$ to contain a large collection of different sectors. In this situation it is convenient for several reasons (e.g. the discussion of collision theory) to enlarge the algebra by adding unobservable elements which connect the different sectors so that $\mathscr{F}$ becomes again a single sector of states for the enlarged algebra $\hat{\mathscr{R}}$. The step from $\mathscr{R}$ to $\hat{\mathscr{R}}$ is somewhat reminiscent of the relation between a group and its covering group. Therefore, I shall call $\hat{\mathscr{R}}$ the covering algebra. The automorphisms of $\hat{\mathscr{R}}$ which leave all elements of $\mathscr{R}$ unchanged are the gauge transformations. It is clear that these matters are intimately connected with the possible statistics of particles. For instance, if a single particle state is in the same sector as the vacuum then the particle is a Boson. It has also been suggested that there is a close connexion between gauge groups and the classification of interactions. However, there are at present more question marks than answers in this whole context, so that it may be better for me to stop at this point. My hope was to sketch for you a few rather deep structural problems of our physical theory which are not far apart from some of the trends in modern mathematics.

References

(1) Doplicher, S., Kastler, D., and Robinson, D. W., *Commun. Math. Phys.* **3**, 1 (1966).
(2) Ruelle, D., *Commun. Math. Phys.* **3**, 133 (1966).
(3) Kastler, D., and Robinson, D. W., *Commun. Math. Phys.* **3**, 151 (1966).
(4) Robinson, D. W., and Ruelle, D., Extremal Invariant States, *Ann. Inst. Henry Poincaré*, to appear.
(5) Doplicher, S., Kadison, R. V., Kastler, D., and Robinson, D. W., *Commun. Math. Phys.* **6**, 101 (1967).

Discussion on the report of R. Haag

A. S. Wightman. I would like to comment on the present state of the problem of proving the existence of solutions for non-trivial models. The models under consideration are super-renormalizable; at the moment no one seems to be strong enough to cope with an infinite number of divergent diagrams. The new results of the past year include a proof by Glimm that the Yukawa interaction in two dimensional space-time has a limit for its dynamics as the ultra-violet cut off goes away (but in a box) and a proof for the $\lambda\phi^4$ theory by Jaffe and Powers, that the box can be removed (if the ultra-violet cutoff is kept). No one has yet succeeded in removing both box and cutoff at the same time.

E. C. G. Sudarshan. Some years ago, Segal showed that the universally invariant states form a simplex (by universally invariant, we mean all transformations on a field, or any other systems of infinite number of degrees of freedom) and is therefore not locally compact. Does the simplex theorem that you stated depend upon a weakening of the group? In particular is it true for any non-compact group on true base space? Is it true for an arbitrary non-compact group?

R. Haag. The theorem holds for any non-compact Lie group which acts on the algebra so that the algebra is asymptotically abelian with respect to this action.

DIFFERENT APPROACHES TO FIELD THEORY, ESPECIALLY QUANTUM ELECTRODYNAMICS

Gunnar Källén

Department of Theoretical Physics, University of Lund,
Lund, Sweden.

1. Introduction

The purpose of this talk is to try to give a survey of various techniques, calculational and otherwise, which have been developed over the years for quantized field theories. We shall not strictly adhere to the historical development of the subject and many important contributions have to be left out in this short summary. Nevertheless, a rough historical order will be followed and the historically oldest techniques of Lagrangian field theory will be discussed first. Further, we shall try to analyse the connexion which exists between two approaches which initially appear to be rather different. Our discussion is admittedly very biased and we sincerely apologize to everybody who feels that other approaches to the same problem have been underemphasized in this summary. Unavoidably, the particular selection which is made in a talk of this kind is heavily influenced by the interest and prejudices of the speaker.

2. The original formulation of Lagrangian field theory

When Lagrangian field theory was originally formulated nearly 40 years ago[1], the language used was essentially the following. Consider a system which is classically described by a number of fields. To be specific, we can think of electrodynamics for spin $\frac{1}{2}$ particles which is described by the electrodynamic potentials $A_\mu(x)$ and the matter field $\psi(x)$. On the classical level, the fields and their interaction are described by a Lagrangian $\mathscr{L}$ which is assumed to be a sum of two terms. The first of these, $\mathscr{L}_0(A, \psi)$, describes the theory without interaction while $\mathscr{L}_1(A, \psi)$ describes the interaction. The two fields A_μ and ψ are then understood as quantum mechanical operators fulfilling canonical commutation relations and not as classical c-numbers. The operator which corresponds to the canonical momentum of one field gives rise to a three-dimensional δ-function when commuted with this field for equal time coordinates. Essentially all other commutators vanish. For Fermi fields anticommutators have to be used instead of commutators. Using standard canonical formalism one then constructs a Hamiltonian $H(A,\psi)$ which is also expressible as a sum of two terms. The first term, $H_0(A, \psi)$, describes the system of free particles while $H_1(A, \psi)$ describes the interaction. The field operators are supposed to act on state vectors $|a\rangle$ in a Hilbert space. As is usual in quantum mechanical problems one is particularly interested in setting up a complete set of state vectors which are eigenstates of the total Hamiltonian H. Therefore, one is looking for the most general solution to the following eigenvalue problem

$$H(A, \psi)|a\rangle = H_0(A, \psi)|a\rangle + H_1(A, \psi)|a\rangle = E_a|a\rangle. \qquad (1)$$

The mathematical problem described by this equation is, as everybody knows, extremely complicated. The first attempts to solve the eigenvalue problem in question went essentially along the following lines. By straightforward and essentially elementary techniques it is possible to introduce a representation in the Hilbert space of the state vectors where the free particle Hamiltonian $H_0(A, \psi)$ is diagonal. Formally, this is the same problem as a field theory with no interaction at all. Let us denote the corresponding state vectors by $|a, 0\rangle$.

$$H_0(A, \psi)|a, 0\rangle = E_a^{(0)}|a, 0\rangle. \qquad (2)$$

The explicit form of this construction is so well-known that it should not be necessary to discuss its details here. The next step is then to *assume* that the eigenvectors appearing in Eqn. (1) can be written as linear combinations of the state vectors in Eqn. (2).

$$|a\rangle = \sum_i C_{ai}|i, 0\rangle. \qquad (3)$$

As is well-known, the state vectors $|a, 0\rangle$ can be classified with the aid of the 'particles' they describe. Evidently, these particles are not the real particles observed in nature but, rather, fictitious particles which would be present if there were no interaction. We shall here adapt the terminology that these particles are called the 'mathematical particles'. In this terminology, Eqn. (3) can be said to be an expansion of the physical states $|a\rangle$ or the 'physical particles' in terms of mathematical particles. Quite clearly, the technique used here is essentially an adaption to the problem encountered in field theory of methods familiar in ordinary quantum mechanics. Further, it might be remarked that in this technique one uses throughout a matrix representation of the operators corresponding to one fixed moment of time, say, $t = 0$. In usual terminology, one is working in the Schrödinger picture.

In spite of the fact that the mathematical method indicated above is formally the same as the standard treatment of most quantum mechanical problems, one finds in actual applications of the method that the structure of the equations is much more complicated than the structure of an ordinary quantum mechanical eigenvalue problem. In elementary quantum mechanics one is used to operators H_0 and H_1 which have a finite norm, i.e. which transform any state vector with a finite norm into another state vector with the same property. This, however, is not the case for the operator $H_1(A, \psi)$ which one writes down in quantum electrodynamics relying on the correspondence with classical theory. Therefore, the very existence of the expansion (3) is questionable from the beginning. Also, using this technique one finds that even if it is possible to work out some elementary problems like bremsstrahlung, pair production, Compton scattering etc. and get answers which are both mathematically reasonable and which agree very well with experimental experience, nevertheless and as soon as one tries to extend the calculations to somewhat more complicated problems like radiative corrections, one encounters serious difficulties. Not only do the answers to calculations of this kind involve divergent sums or, rather, integrals but also the explicit form of these terms is highly ambiguous. Implicitly or explicitly any calculation of this kind introduces a cut off quantity in the sums over the intermediate states which appear because of the expansion technique (3). Normally, this cut off is introduced by brute force in a three-dimensional integration over the momentum of some particle. This has as a consequence that the final result of the computation is not Lorentz invariant but depends on the particular coordinate system in which the cut off has been introduced. This is one— and perhaps the main—reason for the ambiguity just mentioned. Even if it is possible to avoid the ambiguity in many cases, especially the problem of the photon self energy seems to offer a formidable obstacle for this calculation technique. From the point of view of practical applications this is perhaps the most serious problem. However, another point which is very interesting in principle should also be mentioned. Even in some very simple models like a scalar field interacting with a point like source it turns out that the mathematical expansion in Eqn. (3) does not really exist. What actually happens is that all the coefficients c become cut off dependent and in such a way that they all go to zero when the cut off goes to infinity[2]. Nevertheless, the physical state is normalized in such a way that the sum of the absolute squares of all the coefficients c is equal to 1 and independent of the cut off. To describe this situation in broad semi-physical terms, one could perhaps say that the probability of finding any given mathematical state in the expansion of the physical states is arbitrarily small in the limit of no cut-off. However, the total probability of finding any state is exactly equal to one. Clearly, this means that the expansion under consideration is extremely singular. It should be emphasized that this mathematical difficulty seems to have essentially no observable consequence in the specific model just mentioned. Actually, the model is rather trivial and no real scattering exists. Nevertheless, it can be shown that every matrix element of the field operators between physical states exists at least after the elimination of an infinite self energy for the external source system. From this point of view, the renormalized solution of the model exists in spite of the fact that the mathematical expansion (3) is singular also after renormalization. Further, there are good arguments nowadays to ascertain that this phenomenon is not intimately related to the elementary model just mentioned but is very general[3]. The conclusion I would like to draw from this fact is that the mathematical technique which is based on the expansion (3) should be avoided. I want explicitly to mention this point as it appears that a large number of papers has appeared recently where a tremendous mathematical effort is used to describe and discuss the non-existence of the expansion (3)[4]. Actually, in the language I am referring to now, the statement is that the physical states have to be sought in a representation which is *unitary inequivalent* to the representation where the mathematical particles are described. This last representation is sometimes referred to as the 'Fock representation'. Even if a discussion of the relation between the Fock representation and the representation where the physical states are normalizable offers interesting mathematical problems, my personal feeling is that it does not contribute to our understanding of the physics.

3. An alternative formulation of Lagrangian field theory

About twenty years ago we learned something which, nowadays and in retrospect, could be interpreted to mean that the expansion in Eqn. (3) of the physical states in terms of mathematical states is an unnecessary mathematical complication which can be avoided[5]. There are several alternative descriptions of this new technique but I shall here describe only the particular version which I personally like best. In this language one abandons the Schrödinger picture which is basic for the expansion (3) and considers instead the Heisenberg picture where the field operators are dependent on time but where the state vectors are constants. In such a picture there is no Schrödinger equation but, instead, equations of motion for the field operators. For the particular case of quantum electrodynamics which we have in mind here, we shall symbolically write these equations of motion as follows

$$\Box A_\mu(x) = -j_\mu(x), \qquad (4)$$

$$\left(\gamma \frac{\partial}{\partial x} + m\right)\psi(x) = f(x). \qquad (5)$$

The first equation is the Maxwell equation for the electromagnetic potential and the current on the right-hand side is some function of the matter field ψ and, possibly, also of the field A_μ. We do not want to specify the details here as they will not be necessary for our discussion. In a similar way, Eqn. (5) is the equation of motion for the spin $\tfrac{1}{2}$ Dirac field and the operator $f(x)$ on the right-hand side describes the interaction between the matter field and the electromagnetic field. Both the operators $j_\mu(x)$ and $f(x)$ would be zero if there were no interaction. They contain terms with at least one power of e where e is the charge of the electron. If the two differential equations (4) and (5) had been classical equations of motion and if the current $j_\mu(x)$ and the operator $f(x)$ had been known quantities, we should immediately have written down solutions to these two equations in terms of retarded potentials

$$A_\mu(x) = A_\mu^{(\text{in})}(x) + \int D_R(x - x') j_\mu(x')\, dx', \qquad (6)$$

$$\psi(x) = \psi^{(\text{in})}(x) - \int S_R(x - x') f(x')\, dx', \qquad (7)$$

$$\Box D_R(x - x') = \left(\gamma \frac{\partial}{\partial x} + m\right) S_R(x - x') = -\delta(x - x'), \qquad (8)$$

$$D_R(x - x') = S_R(x - x') = 0 \quad \text{for} \quad x_0 < x_0', \qquad (9)$$

$$\Box A_\mu^{(\text{in})}(x) = \left(\gamma \frac{\partial}{\partial x} + m\right)\psi^{(\text{in})}(x) = 0. \qquad (10)$$

The two fundamental retarded solutions $D_R(x - x')$ and $S_R(x - x')$ can be given explicitly in terms of Bessel functions but their detailed form is not interesting for our discussion here. On an intuitive level, the free field solutions $A_\mu^{(\text{in})}(x)$ and $\psi^{(\text{in})}(x)$ describe those parts of the solutions which are not given by the retarded potentials and, therefore, correspond to those fields which were present very long ago before the sources had an opportunity to influence the fields. This intuitive language clearly has as a background the idea that the sources $j_\mu(x)$ and $f(x)$ behave sufficiently regularly and smoothly and vanish sufficiently rapidly in the distant past. In classical theory this is normally the case.

In quantum mechanical applications the source functions are not given functions of space and time but rather functionals of the basic fields themselves. Therefore, the two equations (6) and (7) are not a solution to the quantum mechanical differential equations but just a reformulation of the differential equations in terms of integral equations incorporating a boundary condition at $x_0 = -\infty$. However, the intuitive interpretation is the same, viz. that the solutions to homogeneous equations correspond to fields present at $x_0 = -\infty$ while the retarded potentials describe the influence of the sources on the fields[6]. From the point of view of practical calculations, these two integral equations have the additional feature that the source fields contain at least one power of the coupling constant e. Therefore, the two integral equations (6) and (7) are quite convenient as a starting point for an iterative solution or an expansion in powers of the coupling constant. In this language one would say that in zeroth approximation the Heisenberg fields are equal to the incoming fields. To this we should add a correction term which is obtained as a retarded potential from the source. However, in calculating the correction to the field operator, the source contains one explicit factor e and, therefore, the fields appearing in the source can themselves be replaced by the zeroth order solution, that is, the incoming fields. In higher orders, one sees very easily that one can always be satisfied with an expression of the sources in terms of the fields to order $n - 1$ (and lower) when one wants to calculate the fields to order n. Therefore, the integral equations in principle define the Heisenberg fields as functionals of the incoming fields.

From the historical point of view it could perhaps be remarked that when great progress was made in this area twenty years ago, what we here call the 'incoming fields' were originally termed 'interaction representation fields'. We shall not discuss the historical development here in detail.

Initially, the integration problem as discussed above, i.e. a set of equations which express the Heisenberg fields as functionals of the incoming fields, appears to be a technique entirely unrelated to the eigenvalue problem discussed in the previous section. Nevertheless, the two methods are essentially equivalent. To see how this comes about I should like to make a very short model calculation and discuss the Hamiltonian to first order in the coupling constant e when the iterative solution from Eqns. (6) and (7) is used. To do this we evidently have to specify the interaction Hamiltonian in some detail and write it as follows:

$$H_1(A, \psi) = -\int d^3x A_\mu(x) \frac{ie}{2} [\bar{\psi}(x), \gamma_\mu \psi(x)]. \tag{11}$$

Equations (6) and (7) now give in first order

$$A_\mu(x) = A_\mu^{(\text{in})}(x) + \frac{ie}{2}\int dx' D_R(x-x')[\bar\psi^{(\text{in})}(x'), \gamma_\mu \psi^{(\text{in})}(x')] + \cdots \tag{12a}$$

$$\psi(x) = \psi^{(\text{in})}(x) - ie\int dx' S_R(x-x')\gamma A^{(\text{in})}(x')\psi^{(\text{in})}(x') + \cdots . \tag{12b}$$

We next calculate the total Hamiltonian to this order and find

$$\begin{aligned} H(A, \psi) &\cong H_0(A, \psi) + H_1(A^{(\text{in})}, \psi^{(\text{in})}) = H_0(A^{(\text{in})}, \psi^{(\text{in})}) \\ &+ \frac{ie}{2}\int d^3x \left[\frac{\partial A_\mu^{(\text{in})}}{\partial x_0}\frac{\partial}{\partial x_0} + \frac{\partial A_\mu^{(\text{in})}}{\partial x_k}\frac{\partial}{\partial x_k}\right]\int dx' D_R(x-x') \\ &\times [\bar\psi^{(\text{in})}(x'), \gamma_\mu \psi^{(\text{in})}(x')] \\ &- \frac{ie}{2}\int d^3x \left[\bar\psi^{(\text{in})}(x), \left(\gamma_k \frac{\partial}{\partial x_k} + m\right)\int dx' S_R(x-x')\gamma A^{(\text{in})}(x')\psi^{(\text{in})}(x')\right. \\ &- \frac{ie}{2}\int d^3x\int dx'\left[\bar\psi^{(\text{in})}(x')\gamma A^{(\text{in})}(x')S_A(x'-x)\left(\gamma_k\frac{\partial}{\partial x_k} + m\right), \psi^{(\text{in})}(x)\right] \\ &- \frac{ie}{2}\int d^3x A_\mu^{(\text{in})}(x)[\bar\psi(x), \gamma_\mu \psi(x)] + \text{higher order terms.} \end{aligned} \tag{13}$$

The expression given here looks rather frightening at first sight. However, after a sufficient number of formal partial integration one finds that all the terms to order e actually add up to zero and, therefore, that Eqn. (13) really reads

$$H(A, \psi) = H_0(A^{(\text{in})}, \psi^{(\text{in})}) + \text{terms at least of order } e^2. \tag{14}$$

The formal reason for this at first somewhat astonishing result is, of course, that the Hamiltonian is a constant of the motion. Therefore, the Hamiltonian can be calculated at any time e.g. at $x_0 = -\infty$ which gives just the first term on the left-hand side of Eqn. (14).

Quite evidently, the argument as presented here is extremely formal as several partial integrations are necessary to derive (14) from (13). During these partial integrations one assumes that all surface terms can be left out. Such an argument is quite unreliable even with rather modest requirements on mathematical rigour. Let us for the moment neglect this point but return to it later. If we accept the result (14) it essentially tells us that the complete Hamiltonian expressed as functional of the Heisenberg fields as an operator is identical with the free particle Hamiltonian expressed as a function of the incoming fields. Therefore, the same states which diagonalize the incoming free particle Hamiltonian also diagonalize the complete interacting Hamiltonian and, therefore, are the physical states. If we for the moment allow ourselves to go on with the same kind of intuitive reasoning, we could also say that the canonical commutation relations are independent of time. Therefore, they should formally also hold in the limit $x_0 \to -\infty$, i.e. the incoming fields are also supposed to fulfill the canonical commutation rules. It follows that the problem of setting up a set of states which makes the incoming free particle Hamiltonian diagonal is essentially the same as the corresponding problem for a free field. The solution is well-known and we find that the states labelled by incoming particles are identical with the physical states.

The argument presented here is far from rigorous. To arrive at Eqn. (14) from Eqn. (13) quite a few partial integrations have to be made and all surface terms are left out. Further, the result exhibited in Eqn. (14) is derived only to first order in the coupling constant. To improve both these points, we have to define the behaviour of the interaction terms in a more precise way when the time x_0' approaches minus infinity. A rather brutal but in practice extremely convenient way of doing this is to introduce formally a coupling constant which is weakly time dependent. The general idea is, evidently, that the coupling constant e should have its physical value for all finite times where the interaction is important but that it should vanish sufficiently rapidly for large absolute values of the time. Under these circumstances the partial integrations necessary to go from Eqn. (13) to Eqn. (14) are easily justified. However, and what is more important, relying on this device it is possible to construct an argument which is at least formally independent of the expansion in perturbation theory used here. For this purpose we remark that if the coupling constant is time dependent, energy is not any more exactly conserved because we loose exact invariance under time translations. According to elementary theorems we further know that the change in the energy between two arbitrary times x_0 and x_0' can be expressed in the following way

$$H(x) = H(x_0') + \int_{x_0'}^{x_0} dx_0'' \frac{\partial H(x'')}{\partial e} \frac{de}{dx_0''}, \qquad (15)$$

$$H(x_0) = H[A(x), \psi(x), e(x)]. \qquad (15a)$$

In particular, letting x_0' tend to minus infinity, we have

$$H(x_0) = H_0(A^{(\text{in})}, \psi^{(\text{in})}) + \int_{-\infty}^{x_0} dx_0' \frac{\partial H_1(x_0')}{\partial e} \frac{de}{dx_0'}. \qquad (16)$$

Clearly, Eqn. (16) is the exact counterpart of Eqn. (14) when no expansion technique is used. The general relation (16) also makes it understandable why all the contributions in Eqn. (13) formally add up to zero after partial integrations. To get further, let us assume a special but useful time dependence of the coupling constant e and write

$$e(x_0) = e e^{-\varepsilon|x_0|}, \tag{17}$$

where ε is a very small positive number. In the final result we let ε tend to zero. Next, we remark that a non-diagonal matrix element of Eqn. (16) has a time dependence which is given by the energy difference between the two states.

$$\langle a | \frac{\partial H_1(x_0')}{\partial e} | b \rangle = e^{-i(E_b - E_a)x_0'} \langle a | \frac{\partial H_1}{\partial e} | b \rangle. \tag{18}$$

Equation (18) would be exactly true in a theory invariant under time translations. In our formalism where the Hamiltonian is not exactly time independent, it is to be expected that Eqn. (18) holds in the limit $\varepsilon \to 0$ *assuming that this limit exists*. Under these circumstances we find that the last term of Eqn. (16) becomes

$$\int_{-\infty}^{x} dx_0' \langle a | \frac{\partial H_1(x_0')}{\partial e} | b \rangle \frac{de}{dx_0'}$$

$$= \varepsilon \int_{-\infty}^{x_0} dx_0' e^{-i(E_b - E_a)x_0' - \varepsilon|x_0'|} \langle a | \frac{\partial H_1}{\partial e} | b \rangle$$

$$= \frac{\varepsilon}{\varepsilon + i(E_a - E_b)} \langle a | \frac{\partial H_1(x_0)}{\partial e} | b \rangle e(x) \to 0 \quad \text{for } E_a \neq E_b.$$
(19)

The conclusion to be drawn from this argument is essentially that the non-diagonal matrix elements of the last term on the right-hand side of Eqn. (16) vanish or that the total Hamiltonian including all interactions is diagonal at the same time as the free particle Hamiltonian for the incoming fields. (We disregard here complications implied by the fact that, normally, there are many states with the same energy. This point can be taken care of if we consider other quantum numbers in the formalism like three dimensional momentum, charge etc. Without entering into detail we just state that the final result of such a discussion is as indicated above.) This statement is clearly equivalent to the idea expressed previously that the incoming states are identical with the physical states. Therefore, this integration technique avoids the expansion in Eqn. (3) by constructing the physical states directly with the aid of the incoming field operators and automatically diagonalizes the total Hamiltonian when the equations of motion in Eqns. (6) and (7) are solved.

Admittedly, the argument above is incomplete in one place, viz. it is assumed without further proof that the solutions exist in the 'adiabatic limit' when the parameter ϵ approaches zero and that the solution obtained in that way is invariant against time translations. This is an assumption which is easily verified term by term in a perturbation theory expansion[7] as described above but there is no proof independent of perturbation theory. One more remark should be made, viz. that the expression in Eqn. (19) does not tend to zero in the limit $\epsilon \to 0$ for diagonal terms. This is to be expected as the eigenvalues of the free particle Hamiltonian for the incoming fields and the total Hamiltonian for the Heisenberg fields must be expected to be different even if they are diagonalized simultaneously. If that were not the case, all self masses etc. would be zero. Actually, it is customary to apply a mass renormalization here (as well as some other renormalizations which we shall not discuss in more detail) to ascertain that the two Hamiltonians in Eqn. (16) are not only simultaneously diagonal but also have the same eigenvalues. In a slightly different way of speaking, this means that mass renormalization counter terms are introduced so as to make the mass of the incoming field equal to the mass of the physical Heisenberg field. This is such a well-known procedure nowadays that there is no reason for us to enter upon a detailed discussion here.

Summarizing, this alternative integration technique of Lagrangian field theory uses incoming fields which fulfill the canonical commutation relations. The free particle Hamiltonian calculated from these fields is diagonal at the same time as the complete Hamiltonian and has, after renormalization, the same eigenvalues. This calculation technique has the advantage that it is essentially explicitly Lorentz covariant at any intermediate step. Therefore, divergent expressions of various kinds like self masses etc. can be easily isolated and handled. If the theory were completely finite, the alternative technique described here would still have the advantage over the original formulation of Lagrangian field theory as an eigenvalue problem in the Schrödinger picture that it avoids the very singular expansion (3) of the physical states in terms of mathematical states.

4. Some more recent developments

Even if the language used in the previous two sections is slightly different from the actual language used during the historical development, the material described so far was essentially available in the literature in the early 1950's. It should also be admitted at once that no new development which has contributed significantly to either our calculational techniques or to our general understanding of the subject has really occurred in the last 15 years. Nevertheless, a few things happened which it might be worthwhile describing here. First of all, very many people have realized that a lot of the material which was first known in perturbation theory calculations of quantum electrodynamics has a much more general validity than perturbation theory and can be expressed in terms of what is nowadays known as 'sum rules'. These are relations which express some physical or semiphysical constant inside the formalism as an integral over a spectral function obtained, e.g. from the vacuum expectation value of the product of two or more fields. In the XII Solvay conference a few years ago I gave a short survey of this particular part of the development. I don't think I want to repeat very many of the details here today but I should like to discuss one special problem in this field which has caused much confusion in recent literature. Again, I feel that the basic solution to this problem was given very many years ago but this solution seems to have been forgotten in some more recent publications.

From very general arguments, essentially only Lorentz invariance and certain assumptions about the detailed shape of the mass spectrum of the theory corresponding very closely to what one expects from intuitive reasons, one finds that it is possible to write the vacuum expectation value of a product of two conserved vector fields like, e.g. the currents in quantum electrodynamics in terms of a spectral function $\Pi(p^2)$ in the following way[8]

$$\langle 0| j_\mu(x) j_\nu(x') |0 \rangle = \frac{1}{(2\pi)^3} \int dp e^{ip(x-x')} \Pi(p^2)[p_\mu p_\nu - \delta_{\mu\nu} p^2] \theta(p). \quad (20)$$

The weight function $\Pi(p^2)$ is different from zero only if the vector p is time like ($p^2 < 0$) and is positive definite in that region. The last factor in Eqn. (20), $\theta(p)$, is a step function which is plus one when p_0 is positive and zero otherwise. Further, the weight function can be calculated without any difficulty in first order perturbation theory in quantum electrodynamics and one finds[9]

$$\Pi(p^2) = \frac{e^2}{12\pi^2} \left(1 - \frac{2m^2}{p^2}\right) \sqrt{1 + \frac{4m^2}{p^2}} \, \theta(-p^2 - 4m^2) + 0(e^4). \quad (21)$$

The last factor here is again a step function which says that the weight is zero unless the mass of the vector p is larger than twice the electron mass corresponding to a creation of real pairs. The square root appearing in Eqn. (21) is a phase space factor while the first parenthesis is the result of an explicit calculation using a summation over electron-positron spins.

From very general arguments it is again possible to show that certain moments of the weight function give both the charge renormalization and the self mass of the photon. More in detail, one finds[8]

$$\text{charge renormalization factor} = \int_0^\infty \frac{da}{a} \Pi(-a), \qquad (22a)$$

$$(\text{photon self mass})^2 \sim \int_0^\infty da \Pi(-a). \qquad (22b)$$

The last of these relations is particularly remarkable. We all know very well that the photon self mass has to be identically zero because of gauge invariance. Nevertheless, it is also reasonably well-known for those acquainted with the historical development of the subject that in very many formal calculations a certain, strongly divergent, integral appears in the middle of the computation and has the appearance of being a self mass of the photon. Actually, the divergent integral which one sees in perturbation theory is the expression (22b). Nearly 20 years ago, Pauli and Villars[10] were able to show that a particular cut off technique—since then known as 'regularization'—is able to remove this particular ambiguity in the formalism. Therefore, it is to be expected that a regularization calculation applied to the integral (22b) should give zero in spite of the fact that the integral is strongly divergent and the integrand positive definite. I don't think we will go into details here as they are a little long and published elsewhere[13]. However, one of the characteristic features of the regularization technique is that it is able to eliminate (in spin $\frac{1}{2}$ electrodynamics) some linearly divergent terms while logarithmically divergent terms are not influenced by regularization. As the perturbation theory expression for $\Pi(p^2)$ in Eqn. (21) approaches a constant value for large values of $-p^2$, we see that the integral (22b) is, indeed, linearly divergent. However, if we have a linearly divergent integral of this shape we should in general expect that a logarithmically divergent term also appears. However, if one expands the particular function given in Eqn. (21) in powers of $1/p^2$ one finds that the first term is a constant but that the next term is of order $(1/p^2)^2$. Therefore, there is no logarithmically divergent term in the integral (22b) and regularization, indeed, makes the photon self mass equal to zero also in the language of spectral functions. This is perhaps not unexpected as the regularization procedure was essentially invented to remove the complication of the photon self mass and it should be able to achieve this purpose also in the language of spectral representations.

The more recent interest in this problem comes from the fact that the same divergent integral which has the 'physical' meaning of the photon self mass appears in the equal time commutator of two currents. More in detail, one has formally[11]

$$\langle 0|\ [j_4(x), j_k(x')]\ |0\rangle\bigg|_{x_0=x_0'} = \frac{\partial}{\partial x_k}\delta(\bar{x} - \bar{x}')\int_0^\infty da\ \Pi(-a). \qquad (23)$$

On the other hand, one finds from a formal calculation applying canonical commutation rules to the spin $\frac{1}{2}$ fields that the commutator on the left-hand side of Eqn. (23) should vanish for equal times. However, it should be clear from the remarks made above that if quantum electrodynamics is understood as a limit of the regularized theory, no contradiction appears as the right-hand side is also zero in such an approach. Therefore, and in spite of the rather confused literature which exists about the subject, we feel that no real paradox is involved here, at least not in quantum electrodynamics.

Perhaps it is worth while remarking that the somewhat paradoxical statement made above, viz. that an integral over a positive definite function can effectively be put equal to zero is possible only when we are working with very special and divergent integrals. If we have a case where the integral on the right-hand side of Eqn. (23) is actually convergent, no regularization procedure will ever be able to influence its value. Also, it is quite possible to think of cases where Eqn. (23) does indeed have a non-vanishing right-hand side. One elementary case where that is the situation occurs if one considers a non-interacting vector field and replaces the current operator on the left-hand side of Eqn. (23) by the vector field itself. Formally, the argument leading to the representation (20) still holds and we arrive at Eqn. (23). Further, in that case the weight function is essentially just a δ-function at the mass of the particle and the integral on the right-hand side of Eqn. (23) is identically equal to one. However, that is no contradiction because in such a theory of a divergenceless vector field the time component of the field is not an independent quantity but is essentially equal to the divergence of the canonical momentum which is conjugate to the three space components of the field. As the commutator of the canonical momentum and the field is essentially a δ-function, it follows that the commutator of the time component and the space component of the field is essentially the gradient of a δ-function in complete consistency with Eqn. (23). Also, it is quite clear that the formal operations which are necessary to lead to Eqn. (23) are reliable only when one is working with convergent integrals but one can very easily get contradictions if one makes these formal operations on divergent integrals without considering them as limits of a cut off version of the theory. Finally, we also like to remark that if a relation analogous to Eqn. (23) is applied not to the currents in electrodynamics but to the fields, one finds the rather remarkable result that the vanishing of the right-hand side in that case requires the matrix element between the vacuum and a state with just one (incoming) photon necessarily to be of the form[8]

$$\langle 0| A_\mu(x) |\gamma\rangle = [\delta_{\mu\nu} - Mk_\mu k_\nu] \langle 0| A_\mu^{(in)}(x) |\gamma\rangle, \qquad (24)$$

$$M = \frac{1}{2} \int_0^\infty \frac{da}{a^2} \Pi(-a) = \left(\frac{e^2}{120\pi^2} + \frac{41e^4}{1296\pi^4} + \cdots\right) \frac{1}{m^2}. \qquad (24a)$$

This last result is particularly interesting as it shows that the matrix element in question is not just simply proportional to the corresponding matrix element of the incoming field as one might otherwise be inclined to assume[12].

The techniques with spectral functions and possible sum rules derived from them are evidently not limited to the vacuum expectation value of a product of two operators which we have considered here. However, the attempts to extend this kind of formalism to cases where more than two operators are considered lead to very intricate mathematical problems. Actually, one gets involved with the theory of analytic functions of several complex variables which is much more complicated than the theory of analytic functions of one complex variable. We do not want to enter into these problems here but only state that the actual results of physical interest which have emerged from these investigations so far have been rather limited.

Apart from the development in terms of spectral representations and analytic functions which I have mentioned very briefly here, there has also been another line of research pursued with great vigour during the last ten years or so. I am here thinking of the attempts to describe the whole formalism in terms of field operators averaged only over a finite region in space and time and the algebra constructed from these operators. Actually, I do not feel very competent to talk about this subject but I believe that the problems related to this field will be adequately covered by Professor Haag in his talk. Therefore, I will say no more about them here. Further, I will not speak about attempts which have been made to abandon some of the fundamental postulates of quantum mechanics as examplified, e.g. by the non-linear spinor theory of Heisenberg and collaborators. Also this subject, I believe, will be adequately covered in other talks during this meeting.

References

(1) Heisenberg, W., Pauli, W., *Zs. f. Physik* **56**, 1 (1929); **59**, 168 (1930).
(2) Van Hove, L., *Physica* **18**, 145 (1952). A similar and earlier remark about the same phenomenon but in a different context was made by Friedrich, K. O., *Comm. Appl. Math.* **5**, 349 (1952) and later publications.
(3) E.g. Haag, R., *Dan. Vid. Selsk. Mat. Fys. Medd.* **29**, no 12 (1955); Greenberg, O. W., *Phys. Rev.* **115**, 706 (1959).
(4) Cf., e.g., the summary given by Wightman, A. in the proceedings of the International Physics Conference on Particles and Fields, Rochester 1967 (to be published).
(5) Some of the basic papers are Tomonaga, S., *Progr. Theor. Phys.* **1**, 27 (1946) and later papers; Schwinger, J., *Phys. Rev.* **74**, 1439 (1948) and later papers; Feynman, R., *Phys. Rev.* **76**, 749 (1949) and later papers; Dyson, F. J., *Phys. Rev.* **75**, 486 (1949) and later papers.
(6) Källén, G., *Ark. f. Fysik* **2**, no. 19 (1950) and later papers; Yang, C. N., Feldman, D., *Phys. Rev.* **79**, 972 (1950).
(7) Dyson, F. J., *Phys. Rev.* **82**, 428 (1951).
(8) Umezawa, H., Kamefuchi, S., *Progr. Theor. Phys.* **6**, 543 (1951); Källén, G., *Helv. Phys. Acta* **25**, 417 (1952).
(9) Cf., e.g., Källén, G., *Handbuch der Physik* V_1, Springer 1958.
(10) Pauli, W., Villars, F., *Rev. Mod. Phys.* **21**, 434 (1949).
(11) Goto, T., Imamura, T., *Progr. Theor. Phys.* **14**, 396 (1955); Schwinger, J., *Phys. Rev. Lett.* **3**, 296 (1959).
(12) Lehmann, H., Symanzik K. Zimmermann, W., *Nuovo Cim.* **1**, 205 (1955).
(13) Källén, G., Renormalization Theory in Lectures in Theoretical Physics, Brandeis Summer Institute 1961 (Benjamin, New York 1962).

Discussion on the report of G. Källén

R. Omnes. I would like to bring attention to two recent approaches to field theory and quantum electrodynamics:

(1) The first approach is to use Padé approximants in order to sum the perturbation series. It was shown by Dyson, using physical heuristic argument and it can be explicitly shown on simpler models that the perturbation series is divergent and has an essential singularity at $\alpha = 0$. Recently, Baker and Chrisholm have shown that Padé approximants converge in the case of the Pers model (two coupled harmonic oscillators) which has also an essential singularity at $\alpha = 0$. Bessis and Pusterme have investigated the $\lambda \phi^4$ model. In that case, the Padé approximants automatically ensure unitarity.

(2) In place of treating quantum electrodynamics as an eigenvalue problem or using equations of motion, one can use the analytic properties of the scattering amplitude and compute the Lamb shift, for instance, without meeting any renormalization, and imposing unitarity.

G. Källén. I have nothing to add to your first comment but I would like to disagree with your second point. It is true that if you know exactly where the infinities are in a given theory like electrodynamics, it is possible to arrange the calculations in such a way that you avoid any explicit mentioning of infinities. This has been done, e.g. by Pugh and Rohrlich and many others. However, the infinities are still there and exhibited, e.g. in the high energy behaviour of the weight functions appearing inside dispersion integrals. In the sum rules giving expressions for the self mass etc. in terms of integrals, certain moments of the weights still diverge even if you calculate the weights by dispersion techniques.

W. Heisenberg. In the expansion

$$H(\psi, A) = H_0(\psi^{\text{in}}, A^{\text{in}}) + e(\cdots) + e^2(\cdots) + \cdots$$

which you mentioned, you stated that the term proportional to e vanishes. Does the term proportional to e^2 look like a local or a non-local interaction?

G. Källén. At the first moment it looks very non-local. However, after a sufficient number of partial integrations, many terms add up to zero and the remainder is essentially the self energy term which, in electrodynamics, is local and of the form

$$\delta m = \int \bar{\psi}(x) \psi(x) \, dx$$

M. Levy. I would like to ask two questions about the arguments you gave on the behaviour of the weight function of the vacuum expectation value of the product of two currents:

(1) Do you believe that your argument will remain valid in higher order?

(2) Is the argument also applicable to the product of any two vector currents of $SU(2)$?

G. Källén. The answer to your first question is 'yes, I believe so'. There is a paper by Moffat in the Nuclear Physics several years ago where he claims to make this calculation independently of perturbation theory. In any case, the argument certainly works to order e^4. The answer to your second question is that the argument applies to the total electric current, including both the isovector and the isoscalar part of the hadronic current, but not necessarily to the isovector current alone. I see no reason why the gradient terms should not be there in each term separately.

E. C. G. Sudarshan. In the present report the asymptotic condition did not appear explicitly but nevertheless did come in as an essential step in calculating physical quantities.

Yesterday, Haag stated it is a separate condition; and in his statement the comparison was made between an actual state and a state of many particles 'at infinity'. But with wave packets for particles, when you wait long enough ('at infinity') they do disperse completely. How does one understand the comparison between two states which are both completely dispersed?

G. Källén. If you need a time T for something interesting to happen e.g. for wave packets to disperse, you must choose the parameter ε in such a way that $1/\varepsilon \gg T$. In this way, the 'adiabatic change' in the coupling constant will do you no harm.

R. E. Marshak. Can you really avoid the problem of the indefinite metric with the Pauli–Villars regularization method?

G. Källén. I have nothing new to say at this point. As in the original paper of Pauli and Villars, the indefinite metric is present during this calculation but the states carrying negative probabilities have very large masses and do not influence any physical process at reasonable energies. After the calculation is finished we take the limit where all auxiliary particles get infinitely large masses and thereby disappear from all observable expressions.

F. E. Low. I have a question and a comment. The question is the following: if you consider the function you wrote down for fields rather than currents, then there are two terms. One is of the form you write, $p_\mu p_\nu - \delta_{\mu\nu} p^2$ and is gauge invariant; the second is $p_\mu p_\nu$. I presume you are not claiming that the weight function of the first term is not gauge invariant?

The comment is: in the Van Hove static scalar theory, the overlap integral is equivalent to the usual Z_2 divergence (or zero). You stated that this difficulty existed in all reasonable theories. Now since Z_2 is gauge variant and can be removed, the only divergence in electrodynamics are m and Z_3: now as Johnson, Baker, and Wiley showed, m could well be finite and nobody has shown Z_3 is divergent; therefore your statement is doubtful.

G. Källén. The answer to your question is that the contribution to the weight which is not positive definite comes from the one-photon intermediate states and is, evidently, not gauge invariant. Further, I disagree with your comment. The non-existence of the expansion of the physical states in terms of mathematical states is not equivalent to the infinity of a renormalisation constant, neither Z_2 nor anything else. It is essentially a much weaker statement. I also disagree with your assertion that Johnson and others have shown anything in electrodynamics. Their argument only gives an iteration procedure, the convergence of which is doubtful, to say the least. Further, their explicit mathematics is contradicted by the γ_5 invariance of the model which says that a zero bare mass of the electron implies a zero physical mass of the electron.

F. E. Low. You formed the interaction term diagonal. Does this result essentially rest on the absence of real first order processes?

For example, what would happen with two electrons of different mass which could decay into each other?

G. Källén. My calculation only considers the case of stable particles but should be reasonably general as far as such particles are concerned. I do not know what happens when unstable particles are included.

R. Haag. I think I should make a statement concerning the asymptotic behaviour of the field quantities for large times. No independent assumption about this is necessary (or possible) and the so called 'asymptotic condition' is in fact a relation which follows from the locality structure of the theory. Properly formulated it states that the integral equations written down by Källén are true (for matrix elements) if on the left hand side the wave operators are taken with the physical masses of the particles and if a normalization factor (usually called $Z^{\frac{1}{2}}$) is inserted in front of the incoming fields on the right hand side. With this modification, it is a derivable consequence of locality that the incoming fields satisfy the canonical commutation relations and that

$$H_0(A, \psi) = H_0(A^{\text{in}}, \psi^{\text{in}})$$

Again the physical masses have to be used. Therefore, the term 'asymptotic condition' is misleading and I apologize for having introduced it

12 years ago. To forestall some objections to this statement it should also be emphasized that the statement is not true and in fact the incoming fields are undefinable if the infra-red problem is taken seriously, i.e. if the photon mass is taken to be truly zero. Also the asymptotic relation refers only to the physical (transversal) part of the electromagnetic potential.

H. Umezawa. I myself am also in favour of the method based on the expansion of ψ in terms of in-fields. The question I want to ask here is the following:

When we apply such a method to the Nambu model, we find two different expansions which correspond to the so-called normal and anomalous solutions. In such a case people usually choose one solution among two by comparing the energies of the two ground states. However, there is not much sense in comparing two ground states which respectively belong to different Fock representations. Therefore, I would like to know if there is any clear-cut condition in the field theory which selects a unique solution.

G. Källén. I do not know enough about the model of Nambu to make any comment.

THE NATURE OF PRIMARY INTERACTIONS OF ELEMENTARY PARTICLES*

E. C. G. Sudarshan

Syracuse University, Syracuse, New York

The four-fermion interaction

Over thirty years ago E. Fermi[1] gave a theory of nuclear beta decay in which the neutrino–electron coupling to the nucleons was postulated as a new interaction. Patterned after the structure of quantum electrodynamics, Fermi constructed the fundamental interaction Lagrangian from the scalar product of the four-vector which is formed out of the nucleon spinors with the one formed from the lepton spinors. This is the so-called 'vector' interaction. This led, in the nonrelativistic limit, to spin-independent beta decay interactions. G. Gamow and E. Teller[2] showed that this choice was inadequate to explain the beta decay of the Thorium series, and they proposed a spin-dependent interaction. The differential energy spectrum of the electrons emitted in most beta transitions (the 'allowed' or 'statistical' shape) and the systematic assignment of the reduced lifetimes (the 'ft-values') to various degrees of forbiddenness soon established the essential correctness of the new interaction structure for nuclear beta decay. The succeeding decades saw the discovery of the muon and the processes of muon decay and muon capture by nuclei, and the Fermi interaction described these processes as well. About twenty years ago the outlines of the hypothesis of Universal Fermi Interaction[3] were formulated by O. Klein and G. Puppi, and were elaborated by several authors[4]. However the succeeding years found disparity between the theoretical inferences from beta decay data and the structure of the process of muon decay. It was only after the discovery of parity violation in beta decay that these questions could be resolved. Ten years ago I analyzed the experimental data then existing on weak interactions and I concluded that not all the experiments could be consistent[5]. On the basis of a critical examination of the experiments on beta decay and other weak interactions I was led to the choice of V–A interaction as the only possible universal four-fermion interaction. During the summer of 1957 R. E. Marshak and I developed the concept of chirality invariance as a guiding

* Supported in part by the U.S. Atomic Energy Commission.

principle to search for a systematic derivation of this V–A interaction[6]. We showed that the four-fermion interaction so introduced shared chirality invariance with the canonical commutation relations as well as the electromagnetic interaction, and automatically incorporated the concept of the two-component neutrino. The V–A form disagreed with several experimental results, but these disagreements disappeared when these crucial experiments were repeated during the following year[7]. Feynman and Gell-Mann[8] showed almost immediately that the V–A structure could be inferred from the requirement of a nonderivative coupling within a framework where two-component spinor fields satisfying the Klein–Gordon equation were used. They also showed[8,9] that with the V–A theory it was possible to consider a conserved vector current as the source of the Fermi part of the nuclear beta interaction; in this case the strong nuclear interaction should not lead to any renormalization of the coupling constant for the Fermi type beta interaction[10]. This was very satisfactory since the observed values of the vector coupling constants in muon decay and (Fermi type) nuclear beta decay are practically equal. On the basis of the chiral V–A coupling it is possible to estimate the renormalization of Gamow–Teller beta decay coupling constant and the numerical value so obtained is in good agreement with the observed value[11]. This computation removed the last significant obstacle to the form of the four-fermion interaction that we had proposed for nuclear beta decay[6]. The ten years that have elapsed since this discovery have served to establish it as *the* theory of the weak interaction of nonstrange particles[12].

Need for a new theory

Several new experimental developments in particle physics[13] as well as a desire to find a more fundamental connexion between strong and weak interactions prompt us to reexamine the theoretical basis for the universal four-fermion interaction. Among the weak interaction processes we find that the strangeness violating leptonic decays of hyperons and kaons are slower by a factor of about ten as compared with the estimates made on the basis of a direct strangeness violating four-fermion coupling. There is also unmistakable evidence now for lack of invariance under combined inversion in weak interaction, while the chiral V–A interaction is invariant under combined inversion (though not under charge conjugation or space inversion separately). In the domain of strong interactions a whole collection of vector mesons have been discovered and they seem to play an important role in meson–nucleon and nucleon–nucleon interactions, as well as for understanding the electromagnetic properties of the nucleons. The discovery of the vector mesons by Maglic and coworkers following a method suggested by us and the tentative identification of pseudovector mesons suggests that vector and axial vector fields associated with these particles are related to the vector-axial vector structure of weak interactions. We shall take as a starting point the postulate that the weak and electromagnetic interactions of the nucleon are not primary interactions but are consequences of the weak and electromagnetic interactions of the vector and axial vector meson fields.

Primary interaction Lagrangian

We now propose a theory of primary interactions[14] which retains most of the successes of our chiral V–A interaction[6], but extends it so as to include strange particle decays. It treats the weak and electromagnetic interactions of the strongly interacting particles as secondary interactions, included by virtue of their strong coupling to the vector and axial vector fields. The primary interactions are the direct coupling of the vector and axial vector meson fields with leptons and the photon. The primary interactions are listed below:

(i) *Electromagnetism*

The electrons and muons are coupled to the Maxwell field $\mathscr{A}_\lambda$ according to the standard interaction

$$-e\{\bar{\mu}\gamma_\lambda\mu + \bar{e}\gamma_\lambda e\}\mathscr{A}^\lambda.$$

The nucleons are *not* directly coupled to $\mathscr{A}_\lambda$ but the neutral vector fields ρ_λ and ω_λ are coupled to the Maxwell field according to the linear coupling:

$$-e'\left\{\frac{m_\rho^2}{g}\rho_\lambda + \frac{m_\omega^2}{g}\omega_\lambda\right\}\mathscr{A}^\lambda$$

where g is the strong coupling constant of the vector mesons to the nucleons and, by virtue of the squares of the meson masses, the coupling constant e' is dimensionless. The absolute conservation of electric charge in neutron beta decay implies that

$$e' = e.$$

The vector fields ρ_λ and ω_λ must be divergence-free to ensure the conservation of the electric charge. This implies, in turn, that there are no scalar particles associated with the vector fields ρ_λ, ω_λ.

(ii) *Strong Interactions*

The leptons do not have any strong interactions. The strong interactions involve the Yukawa couplings of the strongly interacting particles with the vector and axial vector fields. The vector fields are divergence-free and therefore describe only vector mesons:

$$\partial^\lambda V_\lambda = 0.$$

The axial vector fields do not have this property, but we may write:

$$A_\lambda = B_\lambda + (\xi/m)\partial_\lambda \phi; \qquad \partial^\lambda B_\lambda = 0;$$

where B_λ describes the pseudovector meson and ϕ describes a pseudoscalar meson of mass m. The dimensionless parameter ξ is a characteristic constant of the theory. The coupling of the (nonstrange) meson fields to the nucleon fields is described by the strong interaction Lagrangian:

$$\tfrac{1}{2}\bar{N}\{g\gamma^\lambda \boldsymbol{\tau}\cdot\boldsymbol{\rho}_\lambda + g'\sigma^{\lambda\nu}\tfrac{1}{2}\boldsymbol{\tau}\cdot\boldsymbol{\rho}_{\lambda\nu} + g_0\gamma^\lambda\omega_\lambda + g'_{00}\sigma^{\lambda\nu}\tfrac{1}{2}\phi_{\lambda\nu}$$
$$+ f\gamma^\lambda\gamma_5\boldsymbol{\tau}\cdot\mathbf{A}_\lambda + f'\sigma^{\lambda\nu}\gamma_5\tfrac{1}{2}\boldsymbol{\tau}\cdot\mathbf{A}_{\lambda\nu} + f_0\gamma^\lambda\gamma_5 E_\lambda + f'_{00}\sigma^{\lambda\nu}\gamma_5\tfrac{1}{2}D_{\lambda\nu}\}N$$

where the symbols stand for the vector and tensor field components of the respective mesons ρ, ω, ϕ, A, E and D.

(iii) *Weak Interactions*

The purely leptonic weak interactions are of the form:

$$\frac{G}{\sqrt{2}}(\bar{e}\gamma_\lambda(1+\gamma_5)\nu_e)(\bar{\mu}\gamma_\lambda(1+\gamma_5)\nu_\mu)^+$$

with possibly a term involving the self-coupling of the $(e\nu_e)$ pair with itself. The nucleons do not have a primary coupling either to the leptons or among themselves. The other primary weak interactions involve the vector and axial vector fields weakly coupled to the baryons or the leptons. The meson–lepton weak interaction is:

$$-\frac{G}{\sqrt{2}}(\bar{e}\gamma_\lambda(1+\gamma_5)\nu_e + \bar{\mu}\gamma_\lambda(1+\gamma_5)\nu_\mu)\times\left(\frac{m_\rho^2}{(g/\sqrt{2})}\rho^\lambda + \frac{m_A^2}{(g/\sqrt{2})}A^\lambda\right).$$

The meson-baryon weak interaction is

$$-\frac{G}{\sqrt{2}}(m_A^2 A_\lambda + m_V^2 \rho_\lambda)J^\lambda$$

where J^λ is a suitable expression which is bilinear in the baryon fields.

Induced electromagnetism

Given these primary couplings we can compute the induced electromagnetic and weak interaction effects. The effective nucleon electromagnetic interaction may be seen to be of the form

$$\frac{e}{2}\bar{N}\{(g_0/g)\Gamma_V(t)\{1-(t/m_\omega^2)\}^{-1}\gamma^\lambda\mathscr{A}_\lambda + \Gamma_V(t)\{1-(t/m_\rho^2)\}^{-1}\gamma^\lambda\mathscr{A}_\lambda$$
$$+ (g'/m_\rho g)\Gamma_T(t)\{1-(t/m_\rho^2)\}^{-1}\tau_3\tfrac{1}{2}\sigma^{\lambda\nu}\mathscr{A}_{\lambda\nu}\}N$$

where $\Gamma_v(t)$, $\Gamma_T(t)$ are the form factors of the vector meson–nucleon vertex. The vanishing of the neutron charge is assured if $g_0 = g$. The nucleon isovector anomalous magnetic moment is given by

$$\mu_1 = (2m_N/m_\rho) \cdot (g'/g) \cdot \Gamma_T(0)$$

The isoscalar magnetic moment vanishes identically. Hence the ratio of the proton and neutron magnetic moments is

$$\mu_P/\mu_n = -\left\{1 + \left(\frac{m_\rho}{m_n}\right)\left(\frac{g}{g'\Gamma_T(0)}\right)\right\}$$

The $SU(4)$ model of strong interactions (see below) gives $\Gamma_T(0)g'/g = 5/3$ and we are then led to predict:

$$\mu_1 = \begin{cases} 4.1 & \text{(theory)} \\ 3.7 & \text{(experiment)} \end{cases}$$

$$\mu_p/\mu_n = \begin{cases} -1.49 & \text{(theory)} \\ -1.46 & \text{(experiment)} \end{cases}$$

which is exceptionally good. The electric and magnetic form factors are given by

$$F_{V,T}(t) = \Gamma_{V,t}(0)\{1 - (t/m_\rho^2)\}^{-1}.$$

If the vector meson vertex falls off in a manner characterized by a single pole structure, the electromagnetic form factors would have dipole structures. This is in agreement with observations[13].

Induced weak interactions

Similar calculations for the effective nucleon beta decay interaction yield:

$$\frac{G}{\sqrt{2}}\bar{p}[\Gamma_V(t)\{1 - (t/m_\rho^2)\}^{-1}\gamma^\lambda + (f/g)\Gamma_A(t)\{1 - (t/m_A^2)\}^{-1}\gamma^\lambda\gamma_5]n$$
$$\times \{\bar{\mu}\gamma_\lambda(1 + \gamma_5)\nu_\mu + \bar{e}\gamma_\lambda(1 + \gamma_5)\nu_e\}.$$

The lack of renormalization of the vector beta coupling and its numerical equality with the muon beta coupling immediately follow. Comparison with beta decay experiments suggests

$$\Gamma_T(0)f/g = g_A \simeq 1.2$$

When we consider the momentum dependent term we obtain the familiar induced pseudoscalar term[15] and the weak magnetism term[9] with the usual *numerical* values.

By virtue of the fundamental principle of the (differential) conservation law of electric charge-current, the neutral components of ρ and ω must remain divergence-free. Hence, we do not expect that they can be coupled to neutral chiral (massive) lepton currents. This is very satisfactory because such neutral lepton currents seem to be totally absent in weak interactions.

It is worth pointing out that the combined inversion (CP) is violated by the 'magnetic' coupling of the axial vector field. This does not lead to any CP violating pion–nucleon coupling, but it does lead to a certain degree of CP violation in nuclear beta decay. But present experiments are not sensitive to this CP violation; for a beta transition with 1 MeV energy release the ratio of the CP violating amplitude to the CP conserving amplitude is about

$$(f'/f)(m_e/m_A) \simeq 10^{-3}.$$

The values of $\Gamma_T(g'/g)$, $\Gamma_A(f/g)$ and $\Gamma_T(f'/g)$ can all be derived from the following line of reasoning. In the low energy limit the vector and axial vector mesons reduce to Fermi type and Gamow–Teller type couplings. The 'electric' coupling of the vector meson is of the Fermi type while the 'magnetic' coupling (i.e., through $\rho_{\lambda\nu}$) is of the Gamow–Teller type. The vector meson couplings may then be identified with the generators of a noninvariance $SU(4)$ group with the nucleon and the $I = J = \frac{3}{2}$ nucleon resonance treated as constituting a representation of the same group[16]. This yields the ratio

$$\Gamma_T(0) \cdot (g'/g) = \tfrac{5}{3}$$

which we have used above. Similarly the low energy limit of the axial vector mesons yield Gamow–Teller interactions for both the axialvector and pseudotensor couplings. We may therefore obtain:

$$\sqrt{\Gamma_A^2 f^2 + \Gamma_T^2 f'^2}/g = \tfrac{5}{3}.$$

If in addition we assume

$$\Gamma_A f \simeq \Gamma_T f'$$

we get

$$\Gamma_A f = \Gamma_T f' = \frac{5}{3\sqrt{2}} g$$

so that

$$g_A = \Gamma_T \cdot (f/g) = 1.2$$

Strange particle weak interaction and the suppression of leptonic decays

For the leptonic decays of the strange particles we extend our theory by the replacement:

$$V_\lambda \to V_\lambda + V'_\lambda$$
$$A_\lambda \to A_\lambda + A'_\lambda$$

where V'_λ and A'_λ are the strange vector and axialvector fields. We shall continue to demand that the vector field remain divergence free and that the longitudinal part of the axial vector field be proportional to the pseudoscalar field:

$$\partial^\lambda V'_\lambda = 0; \qquad \partial^\lambda B'_\lambda = 0.$$

$$A'_\lambda = B'_\lambda + (\xi/m')\partial_\lambda \phi',$$

with the same numerical parameter ξ. We can now calculate the ratio of pion and kaon decay rates into $(\mu\nu_\mu)$ final states:

$$\frac{\Gamma(\kappa \to \mu\nu)}{\Gamma(\pi \to \mu\nu)} = \frac{m_\pi}{m_\kappa} \cdot \left\{\frac{1 - (m_\mu/m_\kappa)^2}{1 - (m_\mu/m_\pi)^2}\right\}^2 \simeq 1.4$$

Using the experimental value of the pion lifetime of 2.55×10^{-8} sec and the branching ratio 0.65 for the two-body leptonic mode of the kaon we predict for the kaon lifetime

$$\tau(K^+) = \begin{cases} 1.17 \times 10^{-8} \text{ sec} & \text{(theory)} \\ 1.22 \times 10^{-8} \text{ sec} & \text{(experiment)} \end{cases}$$

which is in good agreement with the observed value. It is of course essential to note that no new 'smallness parameter' was introduced for describing this interaction.

For the axial vector decays of hyperons we could make use of the relation[17]

$$\partial^\lambda A'_\lambda \simeq \xi m' \phi'$$

It turns out that the effective hyperon Gamow–Teller beta coupling is smaller than the nucleon beta decay coupling by the factor

$$\frac{m_\pi}{m_\kappa}\left(1 + \frac{M - m_n}{2m_n}\right)^{-1} \simeq 0.26,$$

where M is the hyperon mass. On the other hand, as far as the vector beta coupling is concerned, the strange vector meson field has no electric coupling and hence the leading term for the vector beta coupling should vanish. This comes about because a divergence-free vector field cannot couple to two fields with different masses. Hence the vector decay must proceed through the smaller momentum-dependent terms and it is also expected to be significantly suppressed. We see that our theory automatically accounts for the suppression of the strange leptonic decays without any new smallness parameter being introduced.

Consequences for strong interactions

It is possible to make use of the present theory to compute strong interaction effects. Since the calculation involves purely strong interaction phenomena only the lowest order calculations are to be viewed with a certain amount of caution. But it is interesting to observe that we can get good results for s and p wave pion–nucleon scattering lengths by considering nucleon and nucleon resonance exchange and vector meson exchange. The pion–nucleon interaction is not to be postulated anew but is itself an aspect of the axial vector–nucleon coupling; we have the primary pion–nucleon interaction

$$(f_1/m_\pi)\bar{N}\gamma^\lambda\gamma_5\partial^\lambda(\tau\cdot\phi)N;$$

$$f_1 = \tfrac{1}{2}gg_A\xi$$

A simple calculation for the s-wave scattering lengths yields[18] the values (measured in inverse pion masses)

$$\mathscr{A}_1 = \begin{cases} +0.20 \text{ (theory)} \\ +0.183 \text{ (experiment)} \end{cases} \qquad \mathscr{A}_3 = \begin{cases} -0.10 \text{ (theory)} \\ -0.109 \text{ (experiment)} \end{cases}$$

Similarly for p-wave scattering lengths we get

$$a_{11} = \begin{cases} -0.091 & \text{(theory)} \\ -0.101 & \text{(experiment)} \end{cases}$$

$$a_{13} = \begin{cases} -0.022 & \text{(theory)} \\ -0.029 & \text{(experiment)} \end{cases}$$

$$a_{31} = \begin{cases} -0.022 & \text{(theory)} \\ -0.039 & \text{(experiment)} \end{cases}$$

$$a_{33} = \begin{cases} +0.133 & \text{(theory)} \\ +0.215 & \text{(experiment)} \end{cases}$$

With the exception of the resonant channel, these numbers show satisfactory agreement.

In these calculations[18] we have used $g = 9.0$; $\xi = 0.16$. We could relate these parameters to other strong interactions, particularly to meson mass ratios.

Some qualitative features of nucleon–nucleon interaction can be discerned; first of all, the nuclear force consists of three distinct contributions from pseudoscalar, vector and pseudovector particles with their characteristic ranges. The longest range term comes from pion exchange and this 'tail' of the nuclear force has long been known to be consistent with the central and tensor forces coming from pion exchange. Vector meson exchange gives the intermediate range potentials and this includes the leading contribution to the spin-orbit potential The pseudovector meson exchange leads to the shortest range potentials. Since all the coupling constants are now uniquely specified the nucleon–nucleon interaction can be computed.

As an additional strong interaction consequence we know that the nucleon–nucleon potential should not have any r^{-3} singularity at the origin which is known to result if only a pseudoscalar exchange is considered. If we require a cancellation[19] to occur between the π and ρ contributions we are led to a prediction of the ratio of their masses:

$$(m_\pi/m_\rho) = \begin{cases} 0.188 & \text{(theory)} \\ 0.182 & \text{(experiment)} \end{cases}$$

This shows remarkable agreement.

Summary and outlook

We have discussed a theory which attempts to identify specific primary interactions obeying simple rules and leading to a unified treatment of strong, electromagnetic and weak interactions. The vector-axial vector structure, which was first deduced from an analysis of weak interaction data, is seen to be equally important in strong interactions.

The electromagnetic interactions now consist of two types: the familiar 'minimal' interaction of the leptons with the photons, and a different kind of coupling of the neutral vector meson with the photon. Electromagnetism is a primary property of the neutral vector meson, but is only a derived property for the other hadrons. The absence of scalar mesons and the divergence-free nature of the vector field guarantee that the source of the electromagnetic field is always conserved. It is an immediate manifestation of this two-step nature of electromagnetism of nucleons that the nucleon magnetic moments deviate so markedly from those for Dirac particles. A similar mechanism applies for weak interactions also. The suggestion which Yukawa[20] made and later abandoned in connection with meson theory has been resurrected and amplified in the present theory. Beta decay is a derived property of the nucleon but a primary property of the meson. In a manner of speaking the vector and axial vector mesons are intermediate vector bosons for hadron weak interactions. Purely leptonic interactions are direct couplings in the present theory; no intermediate vector bosons are required or expected in muon decay.

The self same difficulties of the local field theory that arise for lepton electrodynamics and for lepton four-fermion interactions continue to be present in this theory. Nor does this theory offer any suggestion as to why there are three categories of interactions. A solution to the problem of divergences of local field theory is outlined in my report on 'Indefinite Metric and Non-local Field Theory' to this Conference.

Such diverse items of particle physics phenomenology as nucleon magnetic moments, beta decay couplings, weak magnetism, absence of neutral lepton currents, absence of scalar mesons, apparent suppression of strange particle decays, pion–nucleon scattering lengths, and the general features of the nucleon–nucleon interaction are correlated in this theory. Such correlations suggest that a fundamental Lagrangian describing the primary interactions is another step forward in our understanding of the study of matter.

References

(1) Fermi, E., *Z. Physik*, **88**, 161 (1934).
(2) Gamow, G. and Teller, E., *Phys. Rev.*, **49**, 895 (1936).
(3) Klein, O., *Nature* **161**, 897 (1948)
    Puppi, G., *Nuovo Cimento*, **5**, 587 (1948).
(4) Tiomno J. and Wheeler, J., *Rev. Mod. Phys.*, **21**, 153 (1949)
    Lee, Rosenbluth and Yang, *Phys. Rev.*, **75**, 905 (1949).
(5) Sudarshan, E. C. G., Doctoral Thesis (University of Rochester 1957).
(6) Sudarshan, E. C. G. and Marshak, R. E., Proceedings of the Conference on Mesons and Newly Discovered Particles (Padua-Venice 1957).
(7) Sudarshan, E. C. G. and Marshak, R. E., *Phys. Rev.*, **109**, 1860 (1958).
(8) Feynman, R. P. and Gell-Mann, M., *Phys. Rev.*, **109**, 193 (1958).
(9) Gell-Mann, M., *Phys., Rev.*, **111**, 362 (1958).
(10) Gerstein, S. S. and Zeldovich, J. B., *JETP*(USSR), **29**, 698 (1955).
(11) Weisberger, W. I., *Phys. Rev. Letters*, **14**, 1047 (1965)
     Adler, S. L., *Phys. Rev. Letters*, **14**, 1051 (1965)
     Tomozawa, Y., *Nuovo Cimento*, **46A**, 707 (1966).
(12) A historical survey of the conceptual developments is given in my contribution to the Centennial Symposium at the American University of Beirut (Ed. Aly, H. H.).
(13) See, for example, Proceedings of the International Conference on Particles and Fields (Rochester 1967).
(14) Sudarshan, E. C. G., *Nature* (in press).
(15) Wolfenstein, L., *Nuovo Cimento*, **8**, 882 (1958)
     Goldberger, M. L. and Treiman, S. B. *Phys. Rev.*, **110**, 1178 (1958).
(16) A review of the relevant concepts is given in Kuriyan, J. G. and Sudarshan, E. C. G., *Phys. Rev.*, **162**, 1650 (1967).
(17) Pradhan, T. and Patnaik, M., Saha Institute (Calcutta) Report (1967) and private communication.
(18) Pradhan, T., Sudarshan E. C. G. and Saxena, R. P., Syracuse University Report (1967); submitted to *Phys. Rev. Letters* for publication.
(19) Rosenfeld, L., Nuclear Forces (North Holland, Amsterdam 1949),
     Schwinger, J. *Phys. Rev.*, **61**, 387 (1942).
(20) Yukawa, H., *Proc. Phys. Math. Soc.*, (Japan) **17**, (1935);
     Rev. Mod. Phys., **21**, 474 (1949).

Discussion on the communication of E. C. G. Sudarshan

S. Weinberg. I would like to ask three questions:

(1) Could you pinpoint just how your calculations differ from those of Schwinger in his paper on 'Partial Symmetry'? There is evidently some difference, because you get a *CP* violation, but it's hard to see just what.

(2) Could you say what general principles underline this sort of calculation? I see that one principle is that things should agree with $SU(4)$ in the non-relativistic limit, but this general requirement seems to be mixed up with ad hoc assumptions.

(3) How do you avoid a large *CP* violation in $K_{\mu 3}$ decay?

E. C. G. Sudarshan. 1. There appear to be several similarities between the strong interaction scheme that I use and the one proposed by Schwinger, particularly in the generality of using both vector and tensor couplings for vector mesons; and in the numerical relation $g_A = 5/3\sqrt{2}$. But I believe the theories are essentially different since I work with second quantized fields and operator Lagrangians, and use therefore the usual machinery of field theory; while Schwinger's formulation uses what he calls 'source theory'. As regards the $g_A = 5/3\sqrt{2}$ relation, initially I thought that my scheme was closely related to Schwinger's scheme, but they cannot be, since he has no *CP* violating terms.

2. With regard to Weinberg's second question I would like to state that there are two parts to my theory. The first part is the statement that hadron electromagnetism and hadron weak interaction are derived properties and occur only by virtue of the coupling of the vector and axial vector mesons with the hadrons. This leads to a certain amount of systematization of various phenomena. This I believe is in a more or less systematic form.

The second part is the arithmetical scheme in which the ratios of the coupling constants are deduced from non-relativistic symmetry principles. This part could stand some more systematization. But we could, in principle, determine the ratio of coupling constants from one set of experimental data and apply it to another.

R. E. Marshak. Weinberg's question about the magnitude of the *CP* violation in *K* decays may not be so difficult to cope with since only the axial vector current requires *CP* violation and this will first enter in K_{e4} where q^2 is not large.

E. C. G. Sudarshan. The *CP* violation introduced in the present theory is in the axial vector coupling but the K_{e3} decay proceeds via the vector coupling. Hence no *CP* violation is expected in K_{e3} decays.

G. Källén. I should like to mention that about a week ago I learned at a meeting in Zagreb that there is a new measurement of the neutron life time from Risø in Denmark.

This new measurement differs from the old determination by about 1 min and replaces $-G_A/G_v = 1.18$ by 1.24 or possibly 1.23. I am not competent to judge which experimental result is correct but I believe there is a possibility that 1.18 has to be changed to a higher value.

N. Cabbibo. The explanation of *CP* violation in this paper coincides, as far as weak interactions are concerned, with an old proposal which introduces *CP* violation through 'second class' currents. If the vector currents (both *CP* even and *CP* odd) are assigned to an octet of currents conserved in the limit of good $SU(3)$ (and this is what Sudarshan is doing), the *T* breaking in $K_{\mu3}$ decay can be proved to be of second order in $SU(3)$ breaking.

Such a theory is not in contradiction with present experimental limits on *T* violation in $K_{\mu3}$.

H. P. Dürr. There is a point which I have not understood completely. Your main argument seems to be that baryons are not coupled directly to the photons but only to vector mesons which in turn are coupled to photons. If one interprets this in the usual sense of a local field theory and tries to reformulate quantum electrodynamics for the baryons using the conventional perturbation procedure one would expect to get all the conventional Feynman diagrams with the difference that now all virtual photon lines are replaced by the product of three propagators with the same 4-momentum, a boson propagator, a photon propagator and again a boson propagator. This case then looks formally similar to the situation where the usual photon propagator is replaced by a regularized photon propagator $1/k^2(k^2 - m^2)^2$, which one would normally get by introducing an indefinite metric in Hilbert space. However, we know that the introduction of an indefinite metric always raises the question whether this theory still leads to a unitary *S*-matrix.

S-MATRIX THEORY WITH REGGE POLES*

Geoffrey F. Chew

Physics Department and Lawrence Radiation Laboratory,
University of California, Berkeley, California

Introduction

The idea of a nuclear theory based directly on the S-matrix goes back to the 1943 papers of Heisenberg[1]. The motivation then, as now, was to avoid unobservable local space–time concepts which become troublesome when quantum principles are combined with relativity. Although macroscopic space–time is an essential component of S-matrix theory, there is no way to construct on-mass-shell wave packets whose spatial localization (in the particle rest frame) is sharper than the particle Compton wavelength. The divergences plaguing conventional local field theory correspondingly have difficulty finding their way into S-matrix theory.

Heisenberg clearly identified two key S-matrix properties, Lorentz invariance and unitarity, and partially appreciated a third—analyticity in momentum variables. The special aspect of the latter that interested him was the correspondence between poles and particles, an essential idea pinpointed by Kramers. S-matrix theorists of the forties, however, did not appreciate the concept now called 'maximal analyticity', and correspondingly they came to believe that interparticle forces are necessarily ambiguous without appeal to local space–time concepts.

During the late fifties the potent dynamical content of analyticity became apparent, with the generalization by Mandelstam[2] of the static pion–nucleon force model that had been developed by Low and myself[3]. Other essential precursors to Mandelstam's double dispersion relations were the single dispersion relations, first formulated correctly for hadrons by Goldberger[4], together with the principle of crossing—identified by Gell-Mann and Goldberger[5]. It is by now familiar how analytic continuation in both angle and energy variables, without any appeal to local space–time, generalizes the Yukawa concept that forces between hadrons are due to exchange of other hadrons.

In studying the dynamics of analyticity by the so-called N/D method, Mandelstam and I in 1960 found a relativistic definition of the force concept and became aware that forces due to exchange of particles considered

* This work was supported in part by the U.S. Atomic Energy Commission.

'composite', like the deuteron or the rho-meson, are unavoidably of the same character as forces due to particles considered 'elementary', like the pion or the nucleon. This led us to the partial 'bootstrap' idea that composite particles may generate themselves, although there was no immediate implication of 'nuclear democracy', i.e. that *all* hadrons are composite. Elementary particles were no longer essential to the dynamics, but they were not excluded. On the other hand it continues today surprisingly often to be forgotten that, once a composite hadron exists, general principles require it to generate forces of a strength and range determined by its mass and partial widths. These forces cannot be 'turned off' at the convenience of theorists in favour of conjectured forces due to exchange of elementary particles.

The definition of the S-matrix in terms of physical observables makes no distinction between elementary and composite particles, but through the pole-particle correspondence may not some difference be found? In his 1961 Solvay report Mandelstam[7] described the attempt to employ a distinction that had arisen in static model work by Castillejo, Dalitz, and Dyson[8], when combined with a theorem of Levinson for potential scattering. The characterization of the so-called CDD poles, however, depends crucially on approximating the unitarity condition by a finite number of channels, no matter how high the energy, while experiments by now seem unequivocal in their indication that dynamics at higher and higher energies is dominated by channels with higher and higher thresholds. The prevailing opinion at present finds the CDD classification of poles unhelpful for the actual hadronic S-matrix.

At the same 1961 Solvay meeting there was for the first time intense discussion of an alternative S-matrix definition of 'compositeness': that nonelementary particles should correspond to poles continuable in angular momentum, i.e. Regge poles. Frautschi and I proposed, as a formulation of 'nuclear democracy', that all hadrons should lie on Regge trajectories[9]. If it were found that certain hadrons did not exhibit the characteristic Regge recurrence at a succession of different spin values, these particles should be classed as elementary. Our suggestion was coupled with the conjecture, still tenable today, that there might exist a unique hadronic S-matrix which not only was Lorentz invariant, unitary, and analytic but which contained only Regge poles. It was proposed in other words that the combination of unitarity, Lorentz invariance and maximal analyticity might suffice to define a 'complete bootstrap' theory of strong interactions, without the need for input parameters or 'master equation of motion'.

This conjecture arose from our inability to discern in models based on master equations any hope for nuclear democracy. A favoured status for certain special quantum numbers seemed inevitable for any master equation within the Lagrangian–Hamiltonian framework; the *absence* of a master equation to us meant 'bootstrap'. You may ask, of course, why we rejected the ancient motion of elementary substructure. This was wishful thinking but with a historical motivation. Suppose some hadronic particle or field should have to be classed 'elementary'; where would that put physics? Back in the morass where it had struggled since the late twenties. The conflict between quantum theory and relativity would remain.

The possibility seemed dazzlingly attractive that, in the hadronic domain, already identified S-matrix principles might render unnecessary the very idea of elementarity. To me at least this possibility was, and is, irresistible.

It is at first exceedingly hard to swallow the notion of dynamics without an equation of motion. Physicists who have learned to live with this apparent absurdity make no pretence of a clear grasp of dynamical completeness for general S-matrix principles. By concrete calculations, however, they have identified a force concept with at least as much precision as in models based on master equations. The difference is that the S-matrix force does not occupy a central position in the dynamics, being merely one link in a circular chain of constraints. In fact the constraints are so severe that no theoretical calculation has come close to satisfying all at the same time. Far from fearing that Lorentz invariance, unitarity and maximal analyticity are insufficient to define a complete dynamical theory, I worry that these requirements may be too much for *any* S-matrix.

In 1961 bootstrappers were insufficiently bold to forego the *a priori* assignment of internal hadronic symmetries, but subsequent bootstrap research, especially by Cutkosky[10], has made it plausible that self-consistency requirements might lead uniquely to the observed pattern of SU_2, SU_3, etc. Arguments also have been given that parity and time reversal may be inevitable attributes of a self-sustaining dynamics.

The programme initiated by Heisenberg two dozen years ago is thus very much alive, the potentialities today appearing more exciting than ever. Let me move now to a more detailed examination of the current position.

Maximal analyticity of the first degree

Although S-matrix theory still lacks a well-defined axiomatic basis, among practitioners there is no divergence of opinion about a set of principles which I shall call 'first-degree analyticity'. One key principle, already implicit in Mandelstam's double dispersion relations[2], states that the only momentum singularities occur at the analytic continuation of the kinematic constraints corresponding to physical multiple processes. By a 'multiple process' I mean a reaction that proceeds via a succession of two or more *macroscopically* separated collisions. For example, double scattering of the type shown in Figure 1 implies a pole in the variable $(p_1 + p_2 - p_4)^2$ at m^2. Triple scattering of the type shown in Figure 2 implies a branch point when the initial and final momenta are such as to

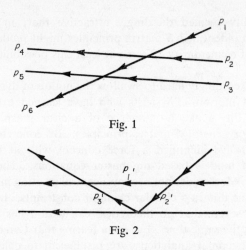

Fig. 1

Fig. 2

allow all three intermediate moments to be physical, that is, on mass-shell and corresponding to macroscopic space–time displacements that correctly add together. Stapp and coworkers have shown the connexion of such singularities with macroscopic causality[11]. It was pointed out by Coleman and Norton[12] that the analytic continuations of these multiple scattering conditions are precisely the Landau rules for the singularities of Feynman graphs. Outside the physical region of real momenta there are delicate questions of sheet structure, but Olive and collaborators[13] have shown how these can be resolved by appeal to the analytic continuation of unitarity, which gives rules for the discontinuity associated with any Landau singularity. Olive's approach shows in fact that the very existence of the Landau singularities could be considered a consequence of unitarity. Macrocausality, in other words, is interlocked with the analyticity–unitarity combination.

Because any multiple scattering may be associated with a graph that looks like a Feynman graph, there is temptation to think of the S-matrix as a superposition of contributions from each of its singularities. The singularities, however, are so intertwined that no simple superposition is possible. For example, a pole typically occurs on only one sheet of that Riemann surface defined by branch points which share the same quantum numbers as the pole. A manifestly correct statement is that sufficiently close to a pole the S-matrix can be approximated through a Laurent expansion by a knowledge of this pole position and residue. Breit-Wigner theory and effective range theory legitimately exploit this circumstance. Physicists nevertheless often may be heard speaking carelessly of a 'pole term'—implying that a given pole makes a well-defined contribution everywhere in the momentum space. There is no basis for the concept 'pole term' in S-matrix theory.

A further source of confusion with Feynman graphs is the factorizability of pole residues and the identification of the individual factors with reaction amplitude (connected parts) of lower dimensionality. The rule, derivable from unitarity, is most conveniently remembered through the multiple scattering diagram. The diagram of Figure 1, for example, not only identifies the pole but gives the physically obvious prescription that each factor in the residue must be a 2 particle $\rightarrow$ 2 particle scattering amplitude. This result sounds like one of the Feynman rules, but the individual factors are themselves complete scattering amplitudes.

By the use of factorization, amplitudes (connected parts) may be unambiguously defined for unstable particles, and the rules for the singularities of unstable-particle amplitudes look like the ordinary rules—except that complex masses occur. By the same token, analysis of unitarity shows that unstable particles must be included among intermediate configurations in the diagrammatic enumeration of singularities. First-degree analyticity ends up treating all particles, stable and unstable, on essentially the same footing, even though the multiple scattering picture at the beginning picks out stable particles for a special role.

Included in what I call 'first-degree analyticity' are the principles of 'crossing' and of 'hermitian analyticity'. Olive[13] has made considerable progress in showing that these properties of the S-matrix are physically inevitable, but their precise axiomatic status remains obscure. Suffice it here to remind you that crossing associates negative energies with antiparticles while hermitian analyticity ensures real masses and coupling constants for stable particles and allows the unitarity condition to be analytically continued as a discontinuity equation.

The above aspects of first-degree analyticity were already embodied in the 1958 Mandelstam representation[2], but that work was restricted to the four-line connected part with low spin. The generalization to arbitrary multiplicity and spin was achieved by Stapp[14] in 1962 with his postulate that it is the M functions, in their dependence on *momentum components*, that have only the Landau singularities. This is a non-trivial point. With more than five lines in a connected part the invariants formed from the momentum vectors are non-linearly related, so the formulation of analyticity properties through invariants becomes ambiguous. So far as spin is concerned, S-matrix elements have Lorentz transformation properties that depend on the momentum variables in a non-analytic fashion. Analyticity properties thus seem to be frame-dependent. Stapp emphasized, however, that the M functions transform in a manner independent of the particle momenta and correspondingly are suitable candidates for a maximal analyticity property. During recent months the importance for practical calculations of the M-function postulate has begun to be widely appreciated.

As stated previously, the validity of the properties included here under the category of first-degree analyticity is at present not an active source of controversy. The logical interrelationship of these properties remains unclear, but in the absence of zero-mass particles there are no signs of inconsistency and the cumulative experimental support is impressive. Zero-mass particles do present an essential complication, requiring a redefinition of the very concept of S-matrix and apparently leading to a totally different singularity structure. The usual approach to electromagnetism exploits the small magnitude of the fine-structure constant by defining an artificial hadronic S-matrix in the absence of electromagnetism and then tacking on the latter by field-theoretical perturbation techniques. Since it appears unlikely that the fine-structure constant and the zero photon mass should be determined by considerations of dynamical self-consistency, this divided approach to strong interactions and electrodynamics is natural when one is thinking in bootstrap terms of a nuclear democracy. The photon, that is to say, has an unmistakably aristocratic appearance.

The leptons are not quite so unique in appearance, but they are very different from hadrons and can hardly be expected to emerge from the same mould. Again, the usual approach is to tack weak-interaction effects by perturbation methods onto the purely strong-interaction S-matrix.

Second-degree analyticity

First-degree analyticity determines all singularities once the poles (i.e. the particles) are given, but a further principle seems required for determining the poles themselves. I say, 'seems', because it has never been shown that first-degree analyticity allows any poles to be arbitrarily assigned. An impression of arbitrariness for spin 0, $\frac{1}{2}$, and 1 is created by renormalizable Lagrangian field theory, which when evaluated by perturbation methods leads order by order to an S-matrix satisfying all the components of first-degree analyticity. The fields and coupling constants in the Lagrangian are, to a considerable extent, arbitrary and thus appear to allow arbitrary assignment of corresponding particles. The flaw in this reasoning is the substantial possibility that power series expansions of the S-matrix inevitably diverge; so it remains conceivable that first-degree analyticity is a sufficient constraint to determine a unique set of particles. Nevertheless, at the present stage of theoretical S-matrix development it is probably unprofitable to be concerned about possible redundancy of assumptions. The first order of business is to find the truth; the most beautiful manner of expressing the truth can wait.

As stated in my introduction, a promising additional S-matrix assumption is that all poles are Regge poles, a principle sometimes designated as 'second-degree analyticity' since it reflects an additional kind of continuability—in angular momentum. The original object of the second-degree assumption was to eliminate the apparent arbitrariness just described as characteristic of renormalizable Lagrangian perturbation field theories. Such field theories have so far always turned out to contain at least one low-spin pole not continuable in angular momentum; so they are excluded by second-degree analyticity. During the past six years, of course, experiments have more and more strongly suggested that established hadrons all lie on Regge trajectories.

Experiments furthermore have tended to confirm the conjecture made independently by Gribov[15] and by Frautschi and Chew[9] that Regge poles control the high energy behaviour of scattering amplitudes at fixed momentum transfers.* This aspect of second degree analyticity has an obvious impact on the question of arbitrary parameters, which in pre-Regge S-matrix theory seemed unavoidable as subtraction constants in dispersion relations. When asymptotic behaviour is interlocked with the poles, arbitrary subtractions become impossible.

Recently, in fact, Horn and Schmid,[16] Logunov, Soloviev and Tavkhelidze,[17] and Igi and Matsuda[18] have extended a technique invented in 1962 by Igi to relate integrals over the low energy resonance region directly to the Regge parameters that control high energy. These Reggeized sum rules† have attractive possibilities for bootstrap investigations, in avoiding

---

* Regge cuts probably exist, as shown by Mandelstam, but the cuts are presumed to be determined by the poles.

† The so-called superconvergence relations may be regarded as special cases of the Reggeized sum rule, applicable when there happen to be no high-lying Regge trajectories or sufficient spin-flip to compensate for high trajectories.

the traditional truncation of unitarity sums that heretofore formed the basis of all dynamical calculations. The relevance to the bootstrap is illustrated in Figure 3, where a few possible high-lying trajectories are sketched; the associated low-energy resonances being indicated. A Reggeized sum rule relates the positive E^2 portion of trajectories belonging

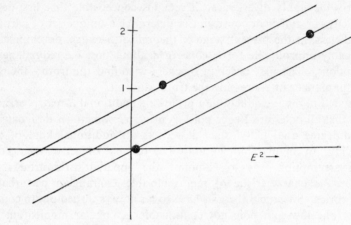

Fig. 3

to a 'direct' reaction to the negative E^2 portion of trajectories belonging to 'crossed' reactions. Evidently the role of direct and crossed reactions may be reversed, the bootstrap problem being to find a set of trajectories that is mutually self-consistent.

Monotonic Regge trajectories

A further and more tentative S-matrix assumption, in this case chiefly motivated by experimental observations, is that the real parts of Regge trajectories as functions of energy squared rise indefinitely as $E^2 \to +\infty$ and fall indefinitely as $E^2 \to -\infty$. A theoretical motivation for this assumption lies in what Fermi used to call 'lack of sufficient reason.' The trajectory asymptotes in dynamical models can always be traced to the presence somewhere in the model of elementary particles. In a complete nuclear democracy it is hard to see where a trajectory would find a reason for going in the complex J-plane—except to infinity.

Mandelstam[19] has probed this question with a perturbation theory model and shown that whereas a trajectory coupled only to 2-particle channels approaches $J = -1$ as $E^2 \to -\infty$, it approaches $J = -2$ when coupled to 3-particle channels, $J = -3$ for 4-particle channels, etc. For positive E^2, model calculations show that trajectories tend to fall after crossing the threshold of the dominant channel. But if higher and higher thresholds tend to dominate as the energy increases, there is no reason for a trajectory ever to reach a maximum. Such reasoning suggests that ultimately it will be unnecessary to assume the monotonic property; this may well turn out to be only self-consistent trajectory behaviour. Current practice, nevertheless, treats monotonicity as an assumption.

It may be imagined that if elementary particles (e.g. quarks) exist, then trajectories will turn over after reaching the elementary particle thresholds and will approach negative-integer asymptotes corresponding to the number of basic particles communicating with the trajectory. One can only guess about such questions, but here is a potential future experimental distinction between an S-matrix governed by the bootstrap and one resting on elementary constituents that are difficult or impossible to observe directly. We obviously need some qualitative evidence for or against elementary particles that can be found in ordinary composite hadron amplitudes. The monotonicity of trajectories may be the kind of feature required. (Notice that if quarks are elementary and are not Regge poles, one could only observe the corresponding fixed asymptotic powers in amplitudes directly involving quarks.)

A related issue is the often-conjectured equivalence of field theory to S-matrix theory. The dynamical content of S-matrix approximations based on unitarity in 'direct' but not in 'crossed' reactions (such as the N/D method), with a finite limit on the particle multiplicity included in the unitarity sum, has so far been expressable through off-shell equations (such as the Bethe-Salpeter equation) in which certain selected particles are given a favoured status. This is roughly equivalent to defining local fields for these particles and seems to support the view that field theory and S-matrix theory are equivalent. With no limit on channel multiplicity, however, and an infinite number of particles to accommodate, it becomes difficult to imagine the form of a Bethe-Salpeter-type equation. Which particles, for example, do you select for special treatment? Perhaps the monotonic-trajectory phenomenon is intrinsically unrepresentable through off-shell equations. If so, it may have an impact on the continuing controversy over the equivalence of field theory to S-matrix theory.

S-matrix models based on monotonic trajectories are beginning to be studied, with Reggeized sum rules expected to provide new insight into the dynamics. One hopes at the same time that the qualitatively encouraging bootstrap indications previously obtained from finite-channel models (either N/D or Bethe-Salpeter), particularly the rough dynamical self-consistency of the observed low-J multiplet structure, will not be lost.

A plausible mechanism to reconcile monotonic trajectories with finite-channel models would be for each physical pole (i.e. particle) to be dominated by the communicating channels of low *kinetic* energy and low *orbital* angular momentum. Moving up along a given trajectory, the dominant communicating channels would thus shift progressively toward those whose constituent particles have appropriately higher-mass and higher-spin. For example, the $\Delta(1240, \frac{3}{2}^+)$ may be dominated (as long believed) by the channel $\pi(140, 0^-)N(940, \frac{1}{2}^+)$, whose threshold is nearest and for which $l = 1$, while the first Δ-trajectory recurrence $\Delta(1920, \frac{7}{2}^+)$ would get a bigger contribution from the channel $\pi(140, 0^-)N(1690, \frac{5}{2}^+)$, at $l = 1$ and with a nearby threshold, than from the channel $\pi(140, 0^-)N(940, \frac{1}{2}^+)$, at $l = 3$ and a distant threshold. To understand an entire trajectory, one needs to consider an infinite number of channels; but to understand *approximately* a small interval along a trajectory, a finite number of channels may suffice.

It is relevant in this connexion to recall that classical nuclear physics from the S-matrix viewpoint is simply the study of baryon number 2 and higher, not to be qualitatively distinguished from high-energy nuclear physics—which may be described as the study of baryon number 0 and 1. A fundamental and successful working principle of classical nuclear dynamics is to consider only those channels whose threshold is close to the particle mass (i.e. nuclear energy level) under consideration.* Needless to remark, the classical nuclear precedent makes it hard to understand, even if quarks exist, how distant-threshold quark-channels could be more important than nearby thresholds in the dynamics underlying low-mass particles.

The mystery of the small pion mass

A number of special correlations and predictions involving the pion have over the years been adduced from field theoretical methods. S-matrix theory has proved in some instances to be less productive than its rival in pionic exploitation, a count not in its favour, and it is important to know whether some essential ingredient is missing or whether the fault lies merely in our inadequately developed understanding of maximal analyticity. As one example, consider the prediction from the so-called 'PCAC hypothesis' that all pion amplitudes should become small near threshold. This prediction follows from the assumption of a smooth off-shell continuation of pion amplitudes that vanishes at zero pion four-momentum: Since the pion mass is small, one is relatively close to the point of zero

---

* In classical nuclear physics the neglected channels are all of higher threshold.

four-momentum whenever the kinetic energy is small. This idea, although imprecise, leads to useful predictions and suggests that, in staying on-shell, S-matrix theory may lack some significant element.

With second-degree analyticity, the pion in S-matrix theory evidently has no special status. There is an infinite number of Regge trajectories, bounded above by the Froissart limit, and one trajectory (not even the highest) happens to cross a physical J-value close to zero energy. (See Figure 4.) Considering the average spacing of trajectories one sees nothing remarkable about this fact. There must be a least massive hadron

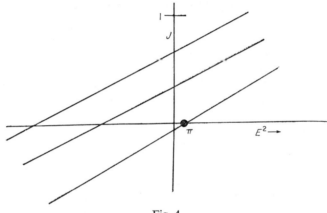

Fig. 4

and it happens to be the pion. Once we know which particle is lightest, is the S-matrix approach capable of yielding especially simple approximate predictions about this particle's properties?

The answer in general is affirmative. Because the pion pole in certain amplitudes comes closer to the physical region than do any other singularities, it is often found that special and simple approximations can be based on pion-pole dominance. This is the phenomenon sometimes called 'peripheralism'. The static model, furthermore, has long exploited the small ratio of pion to baryon mass to yield simple approximate coupling-constant ratios of the type later deduced from group theoretical models. Finally, it has been shown recently that, at zero total energy, Lorentz invariance imposes powerful special constraints on Regge trajectories and residues. It seems likely that such constraints, even though we are at present unsure how to extend them, will have an important effect at a point as close as $E^2 = m_\pi^2 = 0.02$ GeV^2. Superficial arguments suggest a suppressive effect, as in PCAC.

My personal (and perhaps overly optimistic) reading of this picture is that, once given a knowledge of the small pion mass, on-shell analyticity, unitarity, and Lorentz invariance will probably turn out sufficient to produce the successful approximate predictions of off-shell models. A more perplexing issue is the possibility of a direct relation between the small pion mass and the spin and parity, 0^-. Field theory suggests a connexion between an approximately conserved weak axial current and the small mass of a particle whose quantum numbers coincide with the divergence of the axial current. Bootstrap theory, in its exclusion of weak interactions, appears bereft of such a connexion. The low pion mass is supposed to emerge purely from dynamical self-consistency.

A possible resolution of this dilemma is hinted by in a suggestion of Zachariasen and Zweig[20] that the structure of hadron weak interactions could be more complicated than just $J = 1^\pm$ currents. They have shown that $0^\pm$ and higher-spin currents might so far have escaped observation. One could add the conjecture that those special components of the weak hadronic current are enhanced for which corresponding low mass hadrons happen to exist.

A related conjecture is that the apparent locality of weak currents may be an approximation resting on the small mass of appropriate hadrons. It seems conceivable, in other words, that certain equal-time commutation relations are valid to the order of a few per cent because $m_\pi^2/m_N^2 = 0.02$, while other local commutation relations are meaningless because none of the associated hadrons has an especially small mass. Again appealing to classical nuclear physics for a precedent, one may observe that the extremely useful notion of a local nucleon wave function can be derived as an approximation to S-matrix theory but would be untenable if m_π^2/m_N^2 were not small.

If there is something truly fundamental about an approximately conserved weak axial current, then the small pion mass would not seem 'accidental' and the idea of nuclear democracy is in trouble. Note that the SU_3 assignment of the pion to an octet of hadrons, in which all other members have more normal masses, tends to support the 'accidental' interpretation given by a bootstrap-sustained democracy. The unambiguous confirmation of Regge recurrences along the pion trajectory would constitute further support.

The Pomeranchuk trajectory

A second major mystery from the S-matrix standpoint is the possibility of a special status for the vacuum quantum numbers. Striking and distinctive experimental properties attach to the vacuum quantum numbers (hereafter abbreviated as V.Q.N.) which appear to distinguish them from all others. These properties are intimately bound up with the controversial notion of the Pomeranchuk trajectory.

It was conjectured in 1961 by Gribov[15] and independently by Frautschi and Chew[9] that a Regge trajectory belonging to the V.Q.N. (even signature) may pass through the angular momentum value $J = 1$ at precisely zero energy. If other J-singularities are less important, one thereby achieves an immediate understanding of the following five features of high-energy hadron reactions, features that are suggested by experiment:

(1) All hadron total cross sections approach constant nonzero limits at high energy.

(2) All forward elastic amplitudes become pure imaginary in the high energy limit.

(3) For a common target, particle and antiparticle total cross sections approach the same limit.

(4) For a common target, the total cross sections for all members of an isotopic multiplet approach the same limit. (Analogously, to the extent that SU_3 is a 'good' symmetry, the total cross sections for all members of an SU_3 multiplet approach the same limit.)

(5) Those special inelastic reactions where the exchanged quantum numbers are those of the vacuum will dominate at high energy, having the same dependence on energy as elastic scattering.

Properties (3) and (4) had been proposed earlier, before there was experimental evidence, by Pomeranchuk[21,22], using arguments which involved properties (1) and (2). The trajectory supposed to underlie the Pomeranchuk properties was therefore given his name. Property (5) is often called 'diffractive dissociation'.

Once the conjecture had been made of a V.Q.N. trajectory through $J = 1$ at zero energy, predictions became possible for high energy cross sections properties in addition to the above five. The factorizability rule for total cross sections is one of the simplest, but because of the limited variety of hadrons available or targets and beams there are as yet no good experimental tests. A second prediction is that if the slope of the Pomeranchuk trajectory is similar to that of other leading trajectories, there should be an indefinitely continuing shrinkage as the energy increases, in forward peak widths for all reactions, both elastic and inelastic. Unhappily this shrinkage is slow and difficult to observe; for currently accessible energies, similar variations in the shape of the forward peak can be produced by trajectories lying below the Pomeranchuk. Thus we must await the construction of larger accelerators before the existence or nonexistence of asymptotic peak shrinkage can be cleanly established for reactions with V.Q.N. exchange. Even then the effect of branch points in angular momentum may obscure the picture.

The elusive experimental character of elastic peak shrinkage, the most characteristic physical consequence of the conjectured Pomeranchuk trajectory, has permitted widespread scepticism about the existence of this trajectory. A number of trajectories for other quantum numbers are by now regarded as reasonably well established, but severe doubt continues to exist about the V.Q.N. Two different but related sources of scepticism may be pinpointed. First, the dynamical mechanism that would cause the Pomeranchuk trajectory to pass exactly through $J = 1$ at $t = 0$ remains unexplained. Dynamical arguments based on crossing matrices have been given to suggest that a trajectory carrying the V.Q.N. should lie above all others, and the Froissart limit[6] forbids any trajectory from lying *above $J = 1$ at $t = 0$*, but the necessity for an intercept at *precisely $J = 1$* has never been shown. (Remember that no other hadron trajectories pass through integer J at $t = 0$.) Recently it was pointed out[24] that the currently available evidence on the energy dependence of total cross sections does not preclude a Pomeranchuk intercept slightly below 1 (the value $\alpha_P(0) \simeq 0.93$ was proposed), corresponding to total cross sections asymptotically vanishing according to a small negative power of the energy, $\alpha_P(0) - 1$. Should such decreasing behaviour be established, there would be no basis for belief that the V.Q.N. are qualitatively different from other quantum numbers. Most physicists, however, feel it would be ugly for total cross sections to almost, but not quite, approach constants at high energy.

Assuming non-zero limits for high energy total cross sections, a second aspect of the experimental facts seems unnatural from the Regge point of view: The order of magnitude of these hadron total cross-section limits corresponds roughly to the geometrical 'radius' of the particles as defined by the width of the diffraction peak; in other words the cross sections *seem* to be approaching the unitarity limit for orbital angular momenta below that corresponding to the 'radius' as impact parameter. This familiar statement has a simple semiclassical interpretation commonly expressed through optical models, but from the Regge-pole point of view the unitarity limit is not approached; instead the magnitude of a high energy total cross section is determined by a residue of the Pomeranchuk trajectory at $J = 1$. Now remember that at $J = 2, 4, 6$ the Pomeranchuk residues determine the partial widths of mesons with the V.Q.N. (including perhaps the $f(1250)$). Thus, to believe in the Pomeranchuk trajectory one must believe that the analytic extrapolation of a partial width sequence to $J = 1$ is controlled at the latter point by considerations of geometrical 'size'. This idea seems so weird that some physicists dismiss the notion.

What alternative to the Pomeranchuk trajectory is possible that will preserve the properties (1) to (5)? One conceivable alternative is a $J = 1$ V.Q.N. *fixed* pole which has no connexion to physical particles. It was pointed out in 1961 by Froissart[25] that such a fixed pole is allowable if appropriately shielded for $t > 0$ by moving branch points in angular momentum. Gribov[26] had shown earlier that without branch points there would be conflict with unitarity and recently it has been realized that the kind of branch points envisaged by Mandelstam[27] would not suffice: A new type is required, with peculiar but not inconceivable properties.

An objection to a fixed J-pole is the observed qualitative similarity between elastic scattering and inelastic reactions with non-V.Q.N. exchange, such as $\pi^- p \to \pi^0 n$. The order of magnitude of the peak widths is similar, as are the implied pole residues, the larger cross section for the elastic reaction being attributable to the higher V.Q.N. trajectory intercept at $t = 0$. Such phenomenological similarity between elastic and inelastic (non-V.Q.N. exchange) reactions is strange if the underlying mechanism is totally different. It would not be strange if both were due to Regge (moving) poles.

A second objection to a fixed pole at $J = 1$, to be amplified below, is that, with factorization, it would lead to multiple production at high energy in excess of the Froissart limit. It appears, consequently, that if the Pomeranchuk trajectory does not exist we shall require an even more bizarre singularity to replace it.

Thus there continues to be available within S-matrix theory no more palatable an explanation of the special V.Q.N. properties than the bootstrap idea that the highest-lying Regge trajectory carries the V.Q.N. because the attractive forces here are strongest. One important circumstance, however, has never been exploited. This is the distinguished role in the unitarity condition of diagonal S-matrix elements—as opposed to non-diagonal elements. It is well known that under analytic continuation the unit matrix must be separated out and treated on a special basis, and it is precisely the crossed V.Q.N. that communicate with this unit matrix. Here then is an obvious source of exceptional properties for the V.Q.N. The difficulty is that no theorist has so far been clever enough to construct a model satisfying unitarity in both direct and crossed reactions. In consequence we have no understanding of how the poles in a given reaction are affected by unitarity in crossed reactions. One important aspect of this deficiency is our ignorance of the mechanism by which a pole is forbidden from occurring at negative values of energy squared. N/D or Bethe-Salpeter equations fail to preclude such a pole, even though it would violate crossed-unitarity. A second aspect is the Froissart limit forbidding negative-E^2 Regge poles above $J = 1$. This limit is absent from equations that enforce unitarity only in pole-communicating reactions. Crossed unitarity again is essential.

Considering the close relation of such questions to the Pomeranchuk trajectory, it is not surprising that the latter should seem so puzzling. The Pomeranchuk mystery, as opposed to that of the pion, carries no implication of a deficiency in recognized S-matrix principles. The defect lies in our ability to interrelate these principles.

The multi-Regge-pole hypothesis

It was realized in 1963 by Kibble[28] and Ter-Martirosyan[29] that the assumption of Regge asymptotics in one variable of a four-line connected part suggests an extension to several variables in connected parts with more than four lines. In 1965 Toller[30] made a group theoretical analysis of kinematics which can be used to confirm that single-variable Regge asymptotic behaviour, together with factorization, does in fact imply a unique extension to arbitrarily many variables. The extension leads to specific and important predictions about multiparticle production processes, some of which are qualitatively supported by experimental evidence but most of which remain untested. An already mentioned theoretical aspect of the multi-Regge hypothesis is its incompatibility with a fixed $J = 1$ V.Q.N. pole of factorizable residue. Straightforward calculation shows that such a pole would lead to an N-particle production cross section proportional to (log energy)^N, in violation of the Froissart limit[31].

A more positive consequence of multi-Reggeism is the unambiguous definition it implies for connected parts with any number of Reggeized external lines. The Toller analysis, which generalizes a proposal by Joos[32], allows any amplitude to be expanded according to the little groups associated with a set of spacelike momentum transfers, as in Figure 5, which represents a possible momentum-transfer decomposition of a reaction amplitude for four incoming and four outgoing particles. Associated with each momentum transfer Q_i of magnitude $Q_i^2 = t_i$ there is a non-compact sub-group of Lorentz transformations which leaves Q_i unchanged. The amplitude may be expanded in the unitary irreducible representations of this group $[SU(1, 1)]$, which require a continuous label σ. The projection onto a particular representation may be called a 'partial-wave amplitude', which in our example would be designated as $A(\sigma_1, \sigma_2, \sigma_3, \sigma_4)$. If Regge poles exist, corresponding poles would appear in each of the σ variables, the position of a pole in σ_i as a function of t_i constituting the usual trajectory. These poles control the asymptotic behaviour for large elements of the little groups in a manner analogous

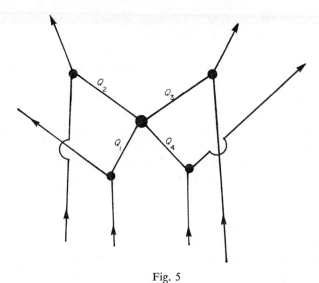

Fig. 5

to that for single Regge poles. By factorization of the pole residues one may then define a connected part each of whose lines is continued in σ, with a simultaneous corresponding continuation in mass. (See Figure 6.)

Although no rigorous connexion has been made between the Joos–Toller partial wave amplitude, defined for spacelike Q, and the Froissart–Gribov amplitude, defined under more restrictive circumstances for timelike Q, Joos, Toller and Boyce[33] have made plausible an analytic relation between the two, with σ recognizable as a complex angular momentum. Thus the $\alpha_i(t_i)$ in Figure 6 may be described as 'complex spins'. Note that we are not going off-shell in these considerations. The mass of each particle follows the spin in such a way that whenever the latter becomes a physical integer of half-integer we are talking about a physical particle.

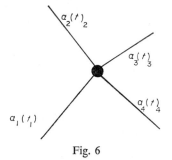

Fig. 6

A conceivable future avenue of dynamical development opens up with the Reggeization of external spins. Heretofore, as I have stressed repeatedly in this report, dynamical models have been limited to a finite number of participating channels. If, however, one somehow could formulate a model in terms of trajectories (rather than individual particles) an infinite number of channels would be included even when the number of trajectories is finite. The approximation scheme presumably would be based on the ranking of trajectories, starting at the top and assuming that the lower the rank the less is the dynamical importance. Such an approach would be very different from the traditional one of keeping only low-mass, low-spin particles in dynamical equations.

Regge daughters and conspiracy

A fascinating recent series of developments has centred on special properties of the S-matrix arising when a momentum transfer Q_i vanishes. The associated little group then enlarges to the full Lorentz group, with a corresponding increase of symmetry. If Regge poles exist, remarkable correlations between *different* poles are required in order to satisfy the increased symmetry. One special aspect of this situation was noticed in 1963 by Gribov and Volkov,[34] and other aspects were found by Domokos and Suranyi[35] in a 1964 Bethe-Salpeter model. The problem has been analysed systematically by Toller[36] on a purely group-theoretical basis.

Toller points out that the unitary irreducible representations of the Lorentz group are labelled by a continuous index λ and a discrete index M. The natural assumption then is that in a $Q_i = 0$ partial wave amplitude $A(\lambda, M)$, poles occur in λ, each pole carrying a definite value of M. These λ poles (called 'Lorentz Poles' by Toller) control $Q_i = 0$ asymptotic behaviour for large group elements in much the same way as do Regge poles for $Q_i \neq 0$. When the larger group $SU(2, 1)$ representations are decomposed in terms of the smaller $SU(1, 1)$ it is found that a single Lorentz pole at $\lambda = \alpha$ corresponds to an infinite series of Regge poles at $j = \alpha - 1, \alpha - 2, \ldots$, with correlated residues and alternating parity and signature. This is the so-called daughter sequence illustrated in Figure 7a. Furthermore, if the Lorentz pole has $M \neq 0$ then Regge poles of *both* parities are required, again with correlated residues, as shown in Figure 7b. This phenomenon has been called 'conspiracy'.

Parity doubling of fermion trajectories at zero energy was pointed out in 1962 by Gribov[37] on the basis of the MacDowell symmetry,[38] but doubling for mesons had not been anticipated. It turns out that the top-ranking meson trajectories have $M = 0$ and consequently are not doubled. There is much discussion and analysis in progress over whether any second-rank

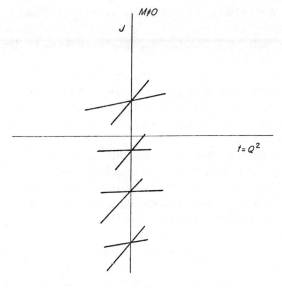

Fig. 7a

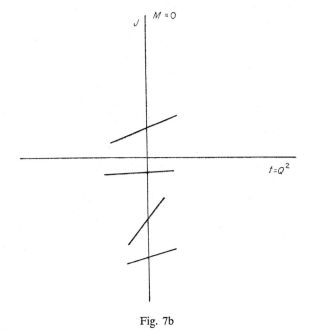

Fig. 7b

trajectories, such as π, B, A_1, etc., may have $M = 1$; the picture at present remains cloudy.

It is remarkable and significant that certain aspects of the above story were independently deduced by arguments that employed analyticity as the primary principle, going not to the point $Q_i = 0$ where the larger symmetry holds but only to the surface $Q_i^2 = 0$. I have already mentioned the phenomenon of parity doubling for fermion trajectories. A second example was the deduction of the daughter sequence by Freedman and Wang[39] from the requirement of compatibility between the Regge expansion and first-degree analyticity in an amplitude for unequal mass particles. The interrelation between analyticity and Lorentz invariance evident in these examples illustrates the difficulty of formulating a non-redundant set of S-matrix axioms.

The deduction from general principles of the daughter sequence reinforces the notion that the number of S-matrix poles must be infinite The new element is that now we anticipate increasing-mass sequences within which all other quantum numbers remain the same.

Conclusion

Let me close this report by emphasizing once again that the S-matrix theory of hadrons may require no further essential physical ingredients. Properties already identified, with solid experimental backing, are more than theorists presently can handle. The problem seems not to be the discovery of additional basic principles but rather the understanding of the mechanism by which recognized properties manage to be mutually compatible. It is here that experiments will continue to play a crucial role. After all, nature has somehow managed to solve the fiercely nonlinear and circular conditions implied by unitarity in different reactions connected by analytic continuation. By looking at enough different aspects of her solution we should be able to figure out how she did it.

A good example of experiment's role in S-matrix theory is the power behaviour at high energy with fixed momentum transfer. Observation of this behaviour has been a powerful stimulant and its implications are far from exhausted. Recently experiments have begun to suggest some kind of exponential behaviour in energy when the angle is fixed. Were such behaviour established as a general phenomenon it would have a major impact. A similar statement evidently applies to the monotonic character of trajectories. One hopes that S-matrix asymptotics ultimately will be deduced from simpler considerations, but experiments may have to lead theorists by the nose to the appropriate mathematical techniques.

Frequently I have alluded to the difficulties of formulating a clean set of S-matrix axioms. This problem seems more and more to be interlocked with the extraction from the axioms of their physical content. A complete statement of first-degree analyticity, for example, may require a definite procedure by which the singularities, including the poles, are systematically located. When and if such a procedure is established it is conceivable that second-degree analyticity will turn out to be redundant, the only poles consistent with unitarity and first-degree analyticity being Regge poles. A related paradox is that traditional axiomatic approaches start with stable particles as given, while in a bootstrap regime such a procedure does not seem natural.

Although one must anticipate periods of theoretical frustation over these enormously difficult questions, we may be confident that the theory will not stagnate so long as experiments are continued. Unless some of the apparently established principles are overthrown, there is present in analytic S-matrix theory a superabundance of still unexploited physical content that cannot be ignored.

References

(1) Heisenberg, W., *Z. Physik*, **120,** 513 and 673 (1943).
(2) Mandelstam, S., *Phys. Rev.*, **112,** 1344 (1958).
(3) Chew, G. F. and Low, F. E., *Phys. Rev.*, **101,** 1570 (1956).
(4) Goldberger, M. L., *Phys. Rev.*, **99,** 979 (1955).
(5) Gell-Mann, M. and Goldberger, M. L., *Proc. 1954 Rochester Conference*, talk by M. L. Goldberger.
(6) Chew, G. F. and Mandelstam, S., *Phys. Rev.*, **119,** 467 (1960).
(7) Mandelstam, S., 'La Théorie Quantique des Champs', *Proc. 12th Solvay Congress*, Interscience, New York (1961).
(8) Castillejo, L., Dalitz, R. and Dyson, F., *Phys. Rev.*, **101,** 453 (1956).
(9) Chew, G. F. and Frautschi, S., *Phys. Rev. Letters*, **7,** 394 (1961).
(10) Cutkosky, R., *Ann. Phys. (N.Y.)*, **23,** 415 (1963).
(11) Stapp, H. P., *Proc. XIIIth International Conference on High Energy Physics*, Berkeley (1966), p. 19. Chandler, C. and Stapp, H. P., UCRL 17734, (1967).
(12) Coleman, S. and Norton, R., *Nuovo Cimento*, **38,** 438 (1965).
(13) Olive, D., *Phys. Rev.*, **135,** B745 (1964); see also Eden, R., Landshoff, P., Olive, D. and Polkinghorne, J., 'The Analytic S-Matrix', Cambridge University Press (1966).
(14) Stapp, H. P., *Phys. Rev.*, **125,** 2139 (1962).
(15) Gribov, V. N., *JETP* **41,** 677 and 1962 (1961).
(16) Horn, D. and Schmid, C., California Institute of Technology preprint, CALT 68-127.
(17) Logunov, A., Soloviev, L. D. and Tavkhelidze, A. N., *Physics Letters*, **24B,** 181 (1967).
(18) Igi, K. and Matsuda, S., *Phys. Rev. Letters*, **18,** 625 (1967).

(19) Mandelstam, S., invited paper at the 1966 International Conference on High Energy Physics, Berkeley (unpublished).
(20) Zachariasen, F. and Zweig, G., *Phys. Rev. Letters*, **14,** 794 (1965).
(21) Pomeranchuk, I. Ya., *JETP* **3,** 306 (1956) and **7,** 499 (1958).
(22) Pomeranchuk, I. Ya. and Okun, L. B., *JETP* **3,** 307 (1956).
(23) Froissart, M., *Phys. Rev.*, **123,** 1053 (1961).
(24) Cabibbo, N., Horwitz, L., Kokedee, J. J. J. and Ne'eman, Y., *Physics Letters*, **22,** 336 (1966).
(25) Froissart, M., invited paper at the 1961 La Jolla Conference on Weak and Strong Interactions (unpublished).
(26) Gribov, V., *Nuc. Phys.*, **22,** 249 (1961).
(27) Mandelstam, S., *Nuovo Cimento*, **30,** 1148 (1963).
(28) Kibble, T. W. B., *Phys. Rev.*, **131,** 2282 (1963).
(29) Ter-Martirosyan, K. A., *JETP* **44,** 341 (1963).
(30) Toller, M., *Nuovo Cimento*, **37,** 631 (1965).
(31) Bali, N. F., Chew, G. F. and Pignotti, A., *Phys. Rev.*, **163,** 1572 (1967); *Phys. Rev. Letters*, **19,** 614 (1967).
(32) Joos, H., 'Lectures in Theoretical Physics', Boulder Summer School (1964) (ed. Brittin, W. E. and Barut, A. O., Boulder, Colo., 1965), Vol. VIIA, p. 132.
(33) Boyce, J. F., International Centre for Theoretical Physics, preprint IC/66/30, Trieste. (April 1966).
(34) Volkov, D. V. and Gribov, V. N., *JETP* **17,** 720 (1963).
(35) Domokos, G. and Suranyi, P., *Nuc. Phys.*, **54,** 529 (1964).
(36) Toller, M., *Internal Reports No. 76 and 84, Instituto di Fisica 'G. Marconi' Roma (April* 1965); CERN preprint TH 780/April 1967).
(37) Gribov, V., *JETP* **16,** 1080 (1963).
(38) MacDowell, S. W., *Phys. Rev.*, **116,** 774 (1960).
(39) Freedman, D. and Wang, J. M., *Phys. Rev. Letters*, **17,** 569 (1966) and *Phys. Rev.*, **153,** 1596 (1967).

Discussion on the report of G. F. Chew

M. Gell-Mann. You have implied that someone who wants to formulate theoretical principles governing the electromagnetic and weak interactions of hadrons and relate them to those of leptons (which seem so similar) would be foolish to introduce local operators: can you give a constructive suggestion as to what we should do instead?

G. F. Chew. I have no immediate suggestion for tackling the problem of currents or, equivalently, form factors, except to employ the standard dispersion relations with an open mind about asymptotic behaviour, which may be related to the question of locality. It seems to me likely that the asymptotic behaviour of the form factor is connected to aspects of S-matrix asymptotic behaviour which we still do not understand, such as the possibility of exponential decrease in certain directions.

J. Hamilton. I wish to ask a practical question, in order to understand the first part of your talk. How am I to calculate the short range part of the interaction in something like the ρ-meson bootstrap, bearing in mind that there is an ellipse of convergence which limits the extent to which crossing can be used to calculate the discontinuity across the unphysical cut of the partial wave amplitude. Can you tell me how to calculate this far away discontinuity, or do you say that the short range part of the interaction is unimportant?

G. F. Chew. I presume that your question about short-range forces is to be understood within the finite-channel N/D framework. As indicated in my report, I feel that this approximation has basic defects that preclude it ever giving a satisfactory answer to the 'short-range' force problem.

The new approach based directly on Regge trajectories recasts the dynamics in such a fundamental fashion that one has not yet identified therein a distinction between force components based on 'range'. This may be a hopeful sign that the old 'short range' dilemma can be circumvented.

J. Hamilton. If that is to be the solution, then it should be emphasized that the method you propose would appear to be considerably removed from the methods used in those applications of dispersion relations which have been successful up to the present.

F. E. Low. How does one define macroscopic causality? i.e. how would one recognize an acausal event that did not violate energy and momentum conservation in some co-ordinate system?

E. P. Wigner. I am afraid I do not have anything very useful to contribute: I do not know of any rigorous definition of causality. As Dr. Low already implied a set of events cannot be said to violate causality ipso facto though it may not be possible to specify which of the events are the causes of the others.

In practice, the situation is different because one uses also some plausibility assumptions. Thus, if two particles collide and then separate, the distance of their separation after a given time cannot exceed a certain value. In order to arrive at a given distance, they must traverse, during the time available, the distance of original separation, and also the distance at which they are finally separated, minus twice some reasonable range of interaction. Expressed in terms of the S matrix, this means that the energy derivative of the phase of the characteristic value must be larger than some negative number, proportional to the 'reasonable range' mentioned before.

However, as the reference to the 'reasonable range of interaction' shows, this is not a sharply defined condition.

B. Ferretti. One way of looking to macroscopic causality is to try to define an asymptotic velocity of signals and to show that this asymptotic velocity has an upper limit which does not exceed the velocity of light *no matter which is the state* of the considered system. This is impossible, unless microcausality is valid, if one considers signals between two space-time points. It should be noted that signals between two space-time points are completely devoided of any physical meaning. However the definition of the asymptotical velocity is possible even if microcausality is not valid, if one considers that the source and the detector of the signals are localised in a finite space-time region of suitable shape, such that a 'frame of reference of the detector' can be defined (see B. Ferretti, *N. Cim.*, **43**, 507, 1966, Figure 2). In this case both a space distance and a time interval with its sign between source and detector can be defined in an invariant way, and so does consequently a speed.

It can then be investigated whether the requirement of the asymptotic upper limit of this speed with increasing distance can be satisfied.

A number of examples in which this requirement is satisfied, in spite of the fact that in these examples the macroscopic causality is not valid, can be constructed (see B. Ferretti, *N. Cim.* **43**, 507, 1966 and *N. Cim.* **43**, 516, 1966).

R. E. Marshak. You state that you hope to develop a complete theory of hadrons and strong interactions on the basis of the bootstrap principle. You do not attempt to understand the electromagnetic and weak interactions of the hadrons ab initio. Your bootstrap method must therefore explain the approximate symmetry groups (broken SU_3, asymptotic $SU_3 \otimes SU_3$ group as reflected in the Weinberg sum rules, etc.) that underlie the hadrons and strong interactions and at the same time yield the $SU_3 \otimes SU_3$ algebra for the hadron currents as well as P.C.A.C. which works so well for the electromagnetic and weak interactions. Is there any chance that the bootstrap method can find this structure without inserting the answers (octets of particles, suitable parity doubling, etc.) at the beginning?

G. F. Chew. There is indeed a chance, in my opinion. I suggest that we wait until Mandelstam has described his recent work before having discussion on this point.

W. Heisenberg. May I make a remark concerning the question of the difference between the bootstrap mechanism and a theory starting from a master-field equation. To my mind, the latter adds only two statements to those of the bootstrap mechanism. It states the group structure of the underlying natural law and it defines the kind of analyticity which should be meant by the postulate of causality. Both statements seem necessary, since I cannot imagine that the bootstrap mechanism really defines the group structure completely (e.g. exact SU_2 or only approximate SU_2), and I wonder whether such terms as maximum analyticity can be defined better than by a differential equation. Would your definition of analyticity differ from that coming from a differential equation?

G. F. Chew. In all Lagrangian field theories with which I am acquainted, the special angular-momentum values selected by the master equation become reflected through a non-democratic particle spectrum. In other words, the theories contain first-degree but not second-degree analyticity.

H. P. Dürr. Chew has mentioned a point which, I think, is important. He indicated that one of the difficulties in writing down a master equation for elementary particles seems to be that it destroys the democracy of elementary particles because it has to be written down in terms of certain fundamental fields which have certain transformation properties and hence will always single out some particles which have the same properties to be more fundamental.

Hence, it appears that writing down a master equation introduces usually more than what Heisenberg has just mentioned, namely certain symmetry group properties and a shorthand description of what we mean by correct analyticity in the S-matrix language arising from the causality requirement, but, in addition, it introduces certain elementary particles, as distinct from composite particles. To establish true democracy for all particles we hence have to learn how to formulate a master equation without introducing an elementary particle. A general prescription to do this is not known. However, it is our impression that by formulating a local field theory in terms of field operators which do not obey the canonical commutation rules we can prevent this field from associating itself directly with an elementary particle and hence avoid this difficulty. At the same time the divergence difficulties can be removed. An alternate method, which probably would achieve the same, would be to introduce canonical fields but with some kind of non-local interactions. From this point of view, it indeed is suggestive to consider the divergence difficulties as being closely connected with the introduction of elementary particles into the formulation. Perhaps somebody knows a better way how to write down field equations without establishing at the same time an elementary particle.

R. Haag. Just a question for information. Is there anything wrong with a fixed pole at angular momentum less or equal to 1?

G. F. Chew. There is no compelling reason to exclude fixed singularities at low J-values (e.g. at $J = \frac{1}{2}$). By definition, however, such singularities would violate the idea of nuclear democracy.

S. Mandelstam. I should like firstly to make some remarks on a dynamical scheme based on rising Regge trajectories. As usual we have the four ingredients

(i) analyticity
(ii) unitarity
(iii) crossing
(iv) bootstrap condition.

As Chew remarked in this talk, the fourth condition is introduced as the requirement that there are no Kronecker-delta singularities in the J-plane rather than as a requirement involving Levinson's theorem.

This last requirement, together with the indefinite rise of Regge trajectories, can easily be imposed, if we work with equations for the Regge trajectories themselves. It is known that one can treat potential theory by using such an approach. The approximation made is that the scattering amplitude is dominated by a finite number of Regge trajectories—in the lowest approximation by one trajectory. The Regge parameters α and β satisfy dispersion relations (this statement is not always true but the complications can be dealt with). Unitarity gives us non-linear equations for the weight functions. The equations were first proposed, I believe, by Zachariasen, and were developed more fully by Cheng and Sharp and others. The dispersion relations for α and β require subtractions; the value of the subtraction terms is determined from knowledge of the potential. The results in the one-trajectory approximation are reasonably accurate for a wide range of Yukawa potentials, especially if modifications of the Regge representation are used.

In the elementary-particle problem one would use two subtractions instead of one in the dispersion integral for α, in order to ensure that the trajectories rise indefinitely. Now the trajectories appear experimentally to be fairly straight lines, which suggests an approximation where one keeps only the two subtraction terms and neglects the dispersion integral entirely. Such an approximation implies infinitely narrow resonances. It simplifies the calculations enormously, non-linear integral equations becoming numerical equations. Furthermore, numerous correlations between resonance parameters have been obtained from current commutation relations or super-convergence relations, and a scheme based on a narrow resonance approximation may well include such correlations. This point of view has been stressed consistently by Gell-Mann. It is one of the advantages of the present scheme that it can be treated in the narrow-resonance approximation. Nevertheless, we can go beyond the narrow-resonance approximation if necessary; we would then have to use equations of the Cheng–Sharp type.

We now come to the question of the subtraction terms in the dispersion integrals. Indeed, in the narrow-resonance approximation, the whole contribution consists of subtraction terms. In the potential model these subtraction terms were introduced from knowledge of the potential; in the elementary-particle problem they must be determined from the crossing relation. There is no unique way of applying the crossing relations, but one attractive possibility is to use the Reggeized sum rules which Chew discussed. He explained how they contain the crossing relation within them.

One has to make several types of approximations. I have already mentioned the narrow-resonance approximation. Another approximation is in the treatment of the Reggeised sum rule, which is only exact if the upper limit of integration is infinite. In practice we have to cut it off above a finite number of resonances and, in the lowest approximation, we cut it off above a single resonance. One does not then expect accurate results, but it is worthwhile to investigate whether the scheme can be implemented to give a consistent, reasonable solution which may serve as a basis for a more adequate treatment.

The particular problem I chose was to obtain the pseudoscalar, vector and axial-vector mesons as bound states of a baryon antibaryon pair. We therefore look at the baryon antibaryon channel, assume that the amplitude is dominated by the three trajectories corresponding to these particles, and cut off the Reggeized sum rules above the lowest resonance. Since we are only investigating one channel we cannot expect to obtain all quantities; the same quantities appear as parameters in different channels. For example, a meson–baryon coupling constant appears in the contribution of a meson resonance to a baryon antibaryon channel, and also in the contribution of a baryon resonance to a meson–baryon channel. We therefore have to take certain quantities from experiment if we look at the baryon antibaryon channel alone. I defined the unit of (mass)² as the inverse of the slope of the trajectory, and took the nucleon to have unit mass on this scale. I also assumed that the vector trajectory was one unit above the pseudo-scalar trajectory and half a unit above the axial-vector trajectory. The equations could then be solved to yield reasonable meson masses ($\mu^2 = 0.3$ for the vector and pseudo-scalar mesons). It also turned out that the ratio of the squares of the coupling constants was positive, a feature that was not automatic from the structure of the equations.

One quantity which could not be calculated from the equations was the magnitude of the coupling constants, as the equations were linear and homogeneous in these variables. To obtain the magnitude of the coupling constant one will have to go beyond the narrow-resonance approximation.

Another bootstrap calculation based on Reggeized sum rules has recently been done by Schmid. He attempted to obtain the ρ as a bound state of the $\pi\pi$ channel. By using the Reggeized sum rules at $t = m_\rho^2$, he was able to avoid making assumptions about the nature of the trajectories. He obtained results in reasonable agreement with experiment.

I should like to make a remark on the subject of the small mass of the pion. It has been shown by Gilman and Harari, following a work by Low, that certain superconvergence relations involving pion scattering are only consistent with saturation by lowest resonances if the pion has zero mass. The saturation by the lowest resonance can clearly not be justified theoretically at present, but it is often assumed and does seem to lead to reasonable results in some cases. If we do assume it we may thus be able to understand the small mass of the pion. In better approximations, where the superconvergence relations are not fully saturated by one resonance, the mass of the pion will not be exactly zero.

I should also like to say something about P.C.A.C. We know that P.C.A.C. imposes certain limitations on the strong interactions. We shall consider the limit where the pion mass is zero; approximate results then become exact. One of the restrictions on the strong interactions is that threshold pion amplitudes vanishes in certain cases (Adler self-consistency condition). Another is that the anti-symmetric part of the amplitude for the scattering of pions off any target is equal to a universal constant multiplied by the isotopic spin of the target. I would like to indicate how these results can be obtained by considering the on-shell strong amplitudes alone, without introducing currents. At the moment I do not have all arguments sufficiently tight, but I believe they are correct.

The assumption that we make is that the conspiracy quantum number M

of the pion trajectory at $t = 0$ is equal to 1. This assumption follows from the following two facts, both of which appear to be true experimentally

(i) The interaction of the pion trajectory with nucleons does not vanish at $t = 0$.

(ii) There is no axial vector particle with mass approximately equal to that of the pion.

It is a feature of conspiracy theory that the ratio of the sense to the nonsense amplitudes is determined by the *kinematics*. In this respect the situation at $t = 0$, where conspiracy theory applies, is different from that at $t \neq 0$, where it does not. Furthermore, with $M = 1$ (or, in general, $M \neq 0$) and with a trajectory passing through $t = 0$ at $J = 0$, which we are assuming, the ratio of the sense to the nonsense amplitudes is zero. The nonsense amplitudes cannot be infinite without violating the required analytic properties in s-t space, so that the sense amplitudes must be zero.

When we examine carefully the implications of the last statement, we realize that they are precisely the Adler self-consistency conditions. If we consider a process $\pi + A \to B$,

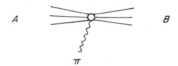

where A and B are in general multi-particle states, we remind the audience that conspiracy theory applies only when all four components of the pion momentum are zero (which implies $m_A = m_B$). Thus, under these conditions, the amplitude vanishes. This is the Adler self-consistency condition.

To proceed further, let us consider an amplitude $\pi_1 + A \to \pi_2 + B$.

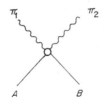

From what we have just said it follows that the amplitude must vanish if either pion has all four momentum components equal to zero. Let us ask the question whether we can find an amplitude which is *linear* in the pion momenta and which satisfies this condition. It turns out that we can find such an amplitude. It has the form

$$A_\mu(p_1 - p_2)_\mu$$

where p_1 and p_2 are respectively the momentum of π_1, and minus the momentum of π_2. A is restricted by the condition

$$A_\mu(p_1 + p_2)_\mu = 0$$

This condition looks very much like the so-called gauge condition in electrodynamics and has similar consequences. For instance, by following Zwanziger's and Weinberg's proof that the interaction constant of a photon with a particle is equal to a universal constant times a conserved quantity, we obtain a similar result here. The only conserved quantity is the isotopic spin, and we thus conclude that the anti-symmetric part of the amplitude for the scattering of pions against a target is proportional to the isotopic spin of the target. This is what we wanted to prove.

Now that we have obtained these on-shell results, let us consider the definition of a current. In the approximation where the pion mass is zero, a partially conserved axial current becomes an exactly conserved axial current. We shall assume that dispersion relations for currents have solutions, though we cannot prove anything in this direction at the moment. If we attempt to construct a conserved axial current, we find poles at $q^2 = 0$, where q is the momentum vector corresponding to the current. It is thus inconsistent to assume the existence of a conserved axial current unless there are psudo-scalar particles with $m = 0$. This is a dispersion-theoretic way of approaching Goldstone's theorem. The coupling of the zero-mass particles must satisfy the Adler self-consistency condition. If we know that we have such particles in the theory, we have no difficulty in constructing a conserved axial current.

Adler and Weinberg have related the commutator between two axial currents to the anti-symmetric part of the amplitude for scattering of pions against a target. We can now reverse their argument and, from our result about this anti-symmetric part, we can show that the commutator between two total axial charges is proportional to the vector charge. If we define the current so that the constant of proportionality be equal to one, and assume weak-interaction universality in the form that the weak Lagrangian involves a current so defined, we obtain the Adler–Weisberger relation in the usual way.

S. Weinberg. Just two short questions to Mandelstam or Chew. Can you calculate the coefficient of proportionality in the relation between pion scattering lengths and the isotopic spin of the target particle? and could you put a 'Reggeized sum rule' on the blackboard?

S. Mandelstam. To reply to the first question: if we assume weak interaction universality in the form stated, we can calculate the scattering lengths in terms of g^2 and $(g_v/g_A)^2$ and obtain the usual Adler–Weisberger formula.

To reply to the second question: if an amplitude has an asymptotic behaviour
$$A(s, t) \cong \sum B_i(t)\{-s\}^{\alpha_i(t)} \qquad s \to \infty$$
then the sum rule takes the form
$$\int^N ds \, \mathrm{Im}\, A(s, t) \sim \sum B_i(t) \frac{N^{\alpha_i(t)+1}}{\alpha_i(t) + s}$$
where the formula is asymptotically true as $N \to \infty$.

R. Omnès. Concerning the problem of the asymptotic behaviour at finite angles mentioned by G. Chew, I want to mention that, at least in the case of $\pi^0 \pi^0$ scattering where crossing is a simple symmetry of the amplitude, falling Regge trajectory contribute to this behaviour an exponentially decreasing expression $e^{-b\theta}$.

E. C. G. Sudarshan. I wish to comment on the bootstrap method of inducing symmetries: in the method pioneered by Cutkosky and followed up by several people it is necessary to put in the correct multiplicity of particles. In this sense a trace of the symmetry is already inserted at the beginning of the bootstrap.

With respect to the Toller expansion in terms of the 0(2, 1) partial waves, I have the following question: since it is known that not all functions can be expanded in terms of these 'partial waves' is one making an assumption? Or has he (or you) proved it?

G. F. Chew. A sufficient condition for the expansion is square integrability, which in Regge terms means that all poles should lie to the left of $J = -\frac{1}{2}$. We believe that such a condition should always obtain for sufficiently large negative values of momentum transfer squared. By analytic continuation from such a region the general partial-wave amplitude may be defined..

A. Tavkhelidze. Question: Were Reggeized sum rules used for bootstrap investigation?

Remark: The reggeized sum rule is a consequence of analyticity and Regge behaviour at infinity. This sum rule connects the low-energy part of the scattering amplitude with high energy behaviour and has been shown to be in a good agreement with experiments.

The requirement that the low energy part be described by resonances alone is an extra condition.

G. F. Chew. This condition seems reasonably well satisfied. Once this additional approximation is invoked, the type of theoretical bootstrap calculation described by Mandelstam becomes possible.

INDEFINITE METRIC AND NONLOCAL FIELD THEORIES

E. C. G. Sudarshan
Physics Department, Syracuse University, Syracuse, New York

Introduction

Quantum theory of fields was invented about forty years ago to provide a proper formulation for electrodynamics but was soon extended to cover electron, meson and nucleon fields. The theory of quantized free fields automatically furnishes symmetric or antisymmetric many particle wave functions according as the quantization is by commutators (as in the case of the radiation field) or by anticommutators (as in the case of the electron field). Indeed, Pauli was able to show that as long as the usual finite-component wave fields were used, consistency of quantization implied the usual connexion between spin and statistics: with symmetric wave functions being associated with integer spin fields and antisymmetric wave functions with half integer spin fields. During recent years it has also been recognized that proper Lorentz invariance for finite-component wave fields implies the so-called TCP invariance. In this scheme the quantized fields are not numerical functions of space and time but operators; or more properly, operator-valued distributions. And it is possible to discuss the theory of quantized free fields in a mathematically satisfactory manner.

The situation is quite different for the theory of interacting quantized fields. For example, in quantum electrodynamics the analogy with classical electrodynamics would suggest an interaction energy density

$$e\,\bar{\psi}(x)\gamma^\lambda\psi(x)A_\lambda(x)$$

where $\psi(x)$ is the spinor field of the electron and $A_\lambda(x)$ is the vector field of the photon, and e is the numerical electric charge of the electron. It is somewhat more satisfactory to consider a properly symmetrized expression

$$\tfrac{1}{2}(\beta\gamma^\lambda)_{rs}[\psi_r^\dagger(x),\,\psi_s(x)]$$

for the current in place of $\bar{\psi}(x)\gamma^\lambda\psi(x)$. If we compute the lowest order in the interaction we recover the results of the unquantized theory including both *stimulated and spontaneous emission*.

* Work supported by the United States Atomic Energy Commission.

This same interaction density however leads to meaningless, formally divergent answers when the effects of the interaction are computed to a higher order. This divergence can be traced back to the local coupling of the electron current and the radiation field and appears in other theories of interacting quantized fields like the pion–nucleon interaction. From a rigorous mathematical point of view the local coupling is not allowed in the form in which it stands; since fields may be thought of as distributions, the local product of such quantities is not in general defined. In other words the interaction density itself is mathematically undefined.

About twenty years ago it was realized that it is possible to extract meaningful answers to a variety of questions involving the interaction of electrons and photons from quantum electrodynamics, answers which are free from infinities provided we express the transition amplitudes in terms of the 'observed' mass of the electron and the 'observed' value of the electric charge. With the aid of the renormalization technique it has been possible to compute the field–theoretic modifications (radiative corrections) to such things as the electron magnetic moment and atomic energy levels. The results obtained are in remarkable agreement with experiment.

Nevertheless there are reasons not to be satisfied with such a quantum theory. First of all, the interaction density that one had started out with is still meaningless and the existence of a field operator for the interacting system is unlikely. Furthermore one can try to compute the mass shifts (i.e. the differences between the observed mass and the mass parameter in the starting Lagrangian) and they turn out to be infinite. The renormalization method, as it stands, is unable to treat other field–theoretic interactions like the chiral V–A interaction responsible for muon decay. Finally, we would like to have a quantum field theory in which the primary fields are themselves suitably defined operators. Within the usual framework of quantized fields which are operator-valued distributions in Hilbert space it does not seem possible to describe a system with a local interaction.

Much effort has been made in recent years to undertake a systematic and mathematically rigorous study of those general features of quantum field theory which are common to all local field theories. The hope has been that in the course of such a study we would learn about the correct method of describing interacting fields. A beautiful theoretical framework has resulted as the by-product of these studies, but the basic question has eluded solution so far. The examples of fields known so far include free fields and certain modifications of free fields like generalized free fields, convex sum of free fields and certain fields which may be though of as polynomials in these fields. It is my understanding that all the results obtained so far are consistent only with free fields (in a suitably generalized sense), being compatible with the rigorous mathematical formulation. In view of this it becomes desirable to study the nature of a quantum field theory with local interaction densities in which we can avoid the divergences which enter the usual formulation.

Even within the renormalization framework it would be desirable to avoid the explicitly divergent quantities. A device of this kind is the method of auxiliary masses or regularization; in this method a formal rule is given to associate a convergent integral depending upon certain auxiliary mass parameters with each divergent integral in the perturbation expansion for any amplitude. The procedure as originally formulated was a covariant method of making divergent integrals finite by an explicitly covariant cut off. It was not derived from the basic formulation of the quantum theory nor does it assure the existence of the interacting field, as a properly defined mathematical entity. We may view regularization as a prescription to substitute a new set of algebraic expressions for the divergent ones deduced from the local interaction theory. It was simply an arbitrary but effective cut off, a veritable bed of Procrustus. A mathematical formulation in terms of fields can be constructed for this invariant regularization scheme by a slight extension of the method of convex combination of fields which we shall outline in a later section, but the field theory so constructed lacks so many of the desirable features for a field theory that it is of no particular interest. We are forced to conclude that the regularization method of Pauli and Villars and of Feynman does *not* provide the basic idea for a new theory of quantized fields.

The regularization method of removing infinities, however, recalls attention to a possible generalization of the mathematical framework of quantum mechanics that was discussed by Dirac several decades ago. This is the possibility of using a linear vector space with an indefinite form for the inner product as the space of state vectors. We shall refer to such a space as an 'indefinite metric space', and to a quantum field theory formulated in such a space as an 'indefinite metric field theory'. In such a theory, we must be careful in the physical interpretation of transition amplitudes as probability amplitudes since physical probabilities should be non-negative. *The proper identification of physical amplitudes is part of the dynamical problem in an indefinite metric quantum theory.*

In the following sections of this report we shall develop a meaningful relativistic quantum field theory with local interactions in an indefinite metric space. This theory yields transition amplitudes which have some superficial resemblance to those obtained by regularization but they are essentially distinct. The mathematical structure of our theory is quite different from that of the field-theoretic version of regularization. In the course of the presentation we shall pose and solve the question of identification of the physical probability amplitudes.

Quantum theory in indefinite metric spaces

In the quantum mechanical description of natural phenomena it has been found very convenient to think of physical states as constituting a vector in a linear space[1]. This is the mathematical expression of the principle of superposition. The correspondence between the physical state and the mathematical vector must be capable of yielding the probability amplitude statements of quantum theory. Therefore, we require that the linear space be equipped with a bilinear inner product $\langle B | A \rangle$ between any two states $|A\rangle$ and $|B\rangle$ which satisfy the following requirements:

$$\langle B | \{c_1 |A_1\rangle + c_2 |A_2\rangle\} = c_1 \langle B | A_1 \rangle + c_2 \langle B | A_2 \rangle \quad \text{(linearity)}$$

$$\langle B | A \rangle = (\langle A | B \rangle)^* \quad \text{(hermiticity)} \tag{1}$$

The probability amplitude for a physical state (corresponding to) $|A\rangle$ being found in the physical state $|B\rangle$ is now identified with the inner product $\langle B | A \rangle$. The probability amplitude is complex but we must demand, as a minimum requirement, that the amplitude for $|A\rangle$ being found in the state $|A\rangle$ has unit absolute value. Hence physical states are normalized; furthermore $|A\rangle$ and $c |A\rangle$ denote the same state if c has unit absolute value. We know that

$$(\langle A | A \rangle)^* = \langle A | A \rangle$$

so that the 'square' of any vector is real. We have now three possibilities; it may be positive, negative or zero. In the conventional scheme of quantum mechanics we further demand that

$$\langle A | A \rangle > 0; \quad |A\rangle \neq 0.$$

In such a case we have an inner product space (over complex numbers) with a positive definite metric. With certain additional mathematical restrictions (completeness and separability) we may identify the physical states as normalized vectors in a Hilbert space. We are interested in exploring the possibility of relaxing this requirement. (If we only consider $\langle A | A \rangle \geq 0$ we get what is sometimes called a pre-Hilbert space. By considering the equivalence classes of all vectors modulo those vectors with zero norm we get a standard Hilbert space). Such a general inner product space is sometimes referred to as an indefinite metric Hilbert space, though it is an unhappy choice of names (especially since the questions of Cauchy sequences etc. are not normally considered). We shall simply refer to them as indefinite metric spaces.

When the states are identified with a linear (definite or indefinite metric) inner product space, the dynamical variables are identified with linear operators in such a space. The lack of simultaneous measurability of two dynamical variables will correspond to the noncommutability of the respective linear operators. The expectation value of a dynamical variable ξ in a state $|A\rangle$ is identified with the quantity:

$$\langle \xi' \rangle = \langle A | \xi | A \rangle$$

provided $|A\rangle$ is normalized. It is more satisfactory to write

$$\langle \xi \rangle = \langle A | \xi | A \rangle / \langle A | A \rangle \tag{2}$$

in which case the state normalization is irrelevant. If the linear operator ξ has as an eigenvector the state $|A\rangle$ with the eigenvalue x we get

$$\langle \xi \rangle = x$$

so that the expectation value and the eigenvalue coincide. 'Real' dynamical variables ought to have only real expectation values. This is assured if the operators are Hermitian (with respect to the inner product): this means that ξ must have the property

$$\langle B | (\xi | A \rangle) = \langle B' | A \rangle^*$$

where

$$|B'\rangle = \xi |B\rangle$$

for any two states $|A\rangle$ and $|B\rangle$. It is then easy to show that ξ has only real expectation values. Let us write

$$|A'\rangle = \xi |A\rangle.$$

Then

$$(\langle A | \xi | A \rangle)^* = (\langle A' | A \rangle)^* = \langle A | A' \rangle = \langle A | \xi | A \rangle \tag{3}$$

so that $\langle \xi \rangle$ is real. This statement is equally valid whether the metric is positive definite or not.

The essential difficulty in using an indefinite metric space is the following: If $|A\rangle$ is any normalized state then $\langle B | A \rangle$ is the probability amplitude that a measurement will find it in the state $|B\rangle$. For a positive metric this

* A unique correspondence between a ket vector $|B\rangle$ and a bra vector $\langle B|$ is presupposed.

amplitude has an absolute value less than unity. But if we have an indefinite metric, for suitably chosen states this amplitude may have an absolute value larger than unity. In this case the probability interpretation breaks down, unless some restriction is placed on which kind of vectors may be made to correspond to physical states. The simplest possibility is to restrict the physical states to be a nonnegative metric (linear) subspace of the larger indefinite metric space. We may refer to these spaces as the 'small' space and the 'large' space respectively. The metric in the 'small' space is to be positive definite. Physical observables must be represented by linear operators in the 'large' space which, however, leave the 'small' space invariant.

If this circumstance obtains the question could be asked: why bother with the large space? The answer to this lies in the fact that all efforts to construct local relativistic field theories describing interactions have been beset with a variety of difficulties which seem to persist as long as the quantized fields are considered as linear operators in a positive definite metric space. One possibility out may be to consider nonlocal relativistic field theories; but it has been found difficult to assure the relativistic invariance of such theories. There is now the option to relax the restriction on the metric: we can consider a local relativistic interacting field theory in a linear inner product space with indefinite metric. The ambiguities and differences which plague field theory can be removed in certain versions of such a theory. We shall show below how to construct a simple Lagrangian field theory and how to identify and isolate the 'small' space of physical states. We may thus answer the question raised above. We want to deal with the 'large' space because it is in this space that we have local relativistic fields[2]. The locality is destroyed when we restrict attention to the small space.

Covariant Lagrangian theory and its diseases

Let us now recapitulate the usual scheme in Lagrangian field theories. We may consider the simple case of a spin $\frac{1}{2}$ (fermion) field of (bare) mass m in interaction with a spin 0 (boson) field of (bare) mass μ. The free Lagrangian density for the system is given by:

$$\mathscr{L} = \bar{\psi}(i\partial\!\!\!/ - m)\psi + \tfrac{1}{2}(\partial^\nu \phi)(\partial_\nu \phi) - \tfrac{1}{2}\mu^2 \phi^2 \tag{4}$$

and the interaction density is

$$\mathscr{H} = g\bar{\psi}\gamma_5\psi\phi. \tag{5}$$

In the interaction picture the operators satisfy free field equations of motion and commutation relations:

$$(i\partial - m)\psi = 0$$
$$(\partial^2 + \mu^2)\phi = 0$$
$$\delta(x^0 - y^0)\{\psi_r^\dagger(x), \psi_s(y)\} = \delta_{rs}\,\delta(x - y) \qquad (6)$$
$$\delta(x^0 - y^0)[\phi(x), \partial^0\phi(y)] = i\delta(x - y)$$
$$\delta(x^0 - y^0)[\phi(x), \psi_r(y)] = 0.$$
$$\delta(x^0 - y^0)[\partial^0\phi(x), \psi_r(y)] = 0$$

The S-matrix is given by the infinite sum:

$$S = 1 + \sum_{n=0}^{\infty} \frac{i^n}{n!} \int d^4x_1 \cdots \int d^4x_n T(\mathscr{H}(x_1) \cdots \mathscr{H}(x_n)) \qquad (7)$$

where T stands for the time ordering operation. By purely mathematical calculations we can, in principle, calculate the scattering matrix from this S-matrix. The technical steps in the covariant calculation involve the use of Wick's rule to convert the time ordered product into a linear combination of normal products with the contraction functions as coefficients. These in turn lead to Feynman's rules for the graphical computation of scattering matrix element using propagators, and enables us to compute quantities like scattering and production amplitudes. But before these computations can be physically interpreted two questions have to be resolved.

The first one involves the fact that the *interaction itself modifies the masses* of the fermion and the boson. These mass shifts themselves can be computed as power series in the coupling constants. We must consider the total Lagrangian to describe particles with this new mass. This is the problem of mass renormalization. It may be restated in the following fashion. Let us consider the mass renormalized free Lagrangian to be

$$\mathscr{L}_0 = \bar{\psi}(i\partial - m_0)\psi + \tfrac{1}{2}(\partial^\nu\phi)(\partial_\nu\phi) - \tfrac{1}{2}\mu_0^2\phi^2 \qquad (4^1)$$

where m_0 and μ_0 are the observed fermion and boson masses, but then the interaction Lagrangian is to be viewed as:

$$\mathscr{H}_0 = g\bar{\psi}\gamma_5\psi\phi + (m_0 - m)\bar{\psi}\psi + \tfrac{1}{2}(\mu_0^2 - \mu^2)\phi^2 \qquad (5^1)$$

and the S-matrix is to be considered as being given by the power series:

$$S_0 = 1 + \sum_{n=1}^{\infty} \frac{i^n}{n!} \int d^4x_1 \cdots \int d^4x_n T(\mathscr{H}_0(x_1) \cdots \mathscr{H}_0(x_n)). \qquad (7^1)$$

This mass renormalization is an essential step in the physical interpretation of the theory and has nothing to do with the infinities which appear in the perturbation expansion[3]. It is to be noted that in this form, the interaction

Lagrangian $\mathscr{H}_0$ is *not* linear in the coupling constant g but the higher order terms are determined by the term linear in g.

The second difficulty is the fact that all except a few lowest order S-matrix elements are *infinite* in each order. This comes about from the divergence of the integrals over the contraction function due to the confluence of the arguments: and thus can be traced directly to the *local* structure of the *interaction* $g\bar{\psi}\gamma_5\psi\phi$. In terms of Feynman's rules these divergences appear as too slow decreases of the product of the various propagators; they are therefore sometimes called 'ultraviolet divergences'. The natural remedy seems to be to consider nonlocal interactions (which would introduce damping of the high frequency contributions at the vertices); or to consider propagators which decrease faster. We shall discuss these remedies in turn.

Nonlocal interactions destroy the foundations of the dynamical theory and of the justification for the equations of motion, unless they continue to be 'local-in-time' i.e. treat space and time asymmetrically[4]. By no means is this logically unsatisfactory, though we would have to verify in detail that, despite this, the theory is truly relativistic. Except in certain very special cases (which we derive by restriction of an indefinite metric theory) no relativistic nonlocal field theory which is local-in-time is known. The suggestion is that the use of indefinite metric to obtain a local finite field theory in the large space which then gets suitably restricted is the most direct method of constructing such theories.

The other alternative of trying to obtain propagators which fall off faster than usual (inverse first power for fermions, inverse second power for bosons) suggests the use of higher order wave equations. Such equations have been studied extensively; and are known to imply an indefinite metric in local Lagrangian field theories. The direct treatment of such a system is in terms of an indefinite metric model of the kind discussed below.

The Lagrangian scheme that we have outlined above and the perturbation expansion are mathematically unsatisfactory in many ways. One can first of all note that the fields are not linear operators but operator-valued distributions; and to get genuine operators we should perform a space average of these fields with suitable testing functions. Once this is recognized, we may continue to take the liberty of dispensing with the explicit reference to such distribution–theoretic refinements. But we have to be careful not to talk of products of field operators at the same space-time point without examining the mathematical existence of such a quantity. For free fields (or for fields in interaction picture) a three-dimensional smearing is enough to produce genuine (unbounded) operators but the products of such operators at the same point do not exist in general. The product $\bar{\psi}(x)\gamma_5\psi(x)$ for example does not exist; consequently, it is not possible to give any precise meaning to the interaction term $g\bar{\psi}(x)\gamma_5\psi(x)\phi(x)$. The perturbation expansion, in its turn, is not mathematically well defined; and so on. As long as we have to take the product of local fields at the same point we would continue to have these mathematically ill-defined quantities, if we work with a positive definite metric space. We shall see below that these mathematical ambiguities would themselves be overcome at the same time as the perturbation theory is made free of infinities by a suitable indefinite metric theory.

Relativistic particle theories: Conflict between covariance and locality

Relativistic quantum theory of particle interactions (without the intermediary of local fields) have been studied for quite some time beginning with the work of Eddington, Peierls, Pryce, Thomas etc.[5] In these theories the primary dynamical variables are *particle variables* and the equations of motion describe the motion of these particles. The formalism of this theory is explicitly noncovariant but nevertheless relativistic. If we use the canonical formalism we may describe this framework as follows: in a relativistically invariant theory there should be ten dynamical variables P_a, J_a, K_a, $\mathcal{H}$, which are the generators of space translations (linear momentum), space rotations (angular momentum), 'boosts' (moment of energy) and time translation (energy). They satisfy the commutation relations of the generators of the Poincaré group. The linear and angular momenta have the same form for free particles and for interacting particles. But, by definition, the Hamiltonian should contain an interaction term. But then the commutation relation

$$[K_a, P_b] = i\delta_{ab}\mathcal{H} \tag{8}$$

shows that the 'boost' generator should also be modified by the interaction. It is then to be expected that the relativistic transformation of the trajectories of particles are no longer manifestly covariant; to the extent that the interaction is nonvanishing the boost transformations also deviate from the transformations for free particles. In other words, (local-in-time) relativistic interacting particle theories cannot be manifestly covariant.*

---

* The local-in-time generator formalism of references 5 and 7 has been further developed by I. Prigogine and collaborators to provide the foundations of relativistic statistical mechanics.

If we give up the notion of a local-in-time dynamical system it is possible to devise relativistic interacting dynamical systems[6]. These systems are, in a certain sense, the analogues of the covariant nonlocal field theories. Probability conservation in such theories seem to be meaningful only asymptotically and the same seems to be true of energy and momentum of the system as well. The unitarity of the S-matrix in such relativistic quantum theories has not been explicitly demonstrated though there is no reason to think that it is not unitary.

But, if we demand both the 'local-in-time' property and manifest covariance it becomes impossible for the theory to have any interaction[7]. Hence, in terms of particle variables it seems impossible to have a relativistic Lagrangian theory for interactions.

It thus appears that the simplest kind of covariant relativistic theory is a local Lagrangian field theory and we must therefore deal with the apparent diseases of such a theory directly. The local-in-time nonlocal relativistic theories are the relativistic version of action-at-a-distance mechanisms characteristic of ordinary particle mechanics. To eliminate this manifest nonlocality and apparent noncovariance we may introduce new dynamical entities like the electromagnetic or meson fields which mediate the interaction. In this fashion action-at-a-distance becomes action-by-contact within the enlarged framework. This enlargement of the theory has proved its merit in both electrodynamics and nuclear interaction; and may be thought of as the realization of Hertz's idea[8] that all forces are due to 'concealed motions' and 'concealed entities' (but within a relativistic quantum framework!). While in many cases the new dynamical entities by themselves do not demand a change in the metric of the linear space, but the Coulomb interaction between charged fields does demand that a manifestly covariant local relativistic form inevitably leads to an indefinite metric for the linear space[9]. The nonlocality that we are most interested in is one somewhat unrelated to this: it is a nonlocality dictated by the need to remove ultraviolet divergences. In a manner of speaking, just as Yukawa's hypothesis introduced new dynamical entities to eliminate the 'form factor' of nuclear collisions (i.e. the nuclear potential), here we introduce new dynamical entities by enlarging the small (physical) space so as to eliminate both the ultraviolet divergences and any nonlocal 'form factors' of the field theory.

Since the time Dirac introduced the indefinite metric into the space of quantum states[10] there have been various cases of field theories and models where second quantization has involved the use of an indefinite metric space. An example which is not always recognized as such is the regularization by the aid of auxiliary masses and fields[11]. While regularization is sometimes viewed as a mathematical technique to handle the ultraviolet divergences, to place it on a Lagrangian basis and even to restore unitarity into the perturbation expansion, quantization requires the use of auxiliary fields, at least one of which satisfies commutation relations with the 'wrong' sign. This, in turn, implies the use of an indefinite metric space. Higher order wave equations of the kind considered by Green and by Pais and Uhlenbeck also introduce an indefinite metric[12]. Finally Heisenberg's unified theory of elementary particles[13] is formulated in an indefinite metric space[14].

The usual objection to an indefinite metric theory is that in view of the appearance of negative probabilities a consistent reinterpretation would actually lead to an elimination of the indefinite metric as far as physical states are concerned. The point of view that we shall pursue is that an indefinite metric theory with local interactions may be the simplest and most elegant method of constructing a relativistic quantum theory that is equivalent to a nonlocal relativistic theory with a positive definite metric. *Manifest covariance for interacting fields makes an indefinite metric inevitable*, but this prospect is not viewed with alarm. A general invariant method of introducing a subsidiary condition is outlined which defines the 'small' physical subspace. In general these subspaces are positive norm eigenstates of the S-matrix.

Covariant field theory in an indefinite metric space

Reconsider the simple model of a fermion field ψ in interaction with a pseudoscalar field ϕ. We found that in the interaction terms $g\bar{\psi}\gamma_5\psi\phi$ the quantity $\bar{\psi}\gamma_5\psi$ was ill-defined. We now extend this model by writing

$$\Psi = c_1\psi^{(1)} + c_2\psi^{(2)} + c_3\psi^{(3)} \tag{9}$$

and

$$\delta(x^0 - y^0)\{\psi_r^{(j)\dagger}(x), \psi_s^{(k)}(y)\} = \eta^{jk}\,\delta_{rs}\,\delta(x - y). \tag{10}$$

We choose the matrix η^{jk} to be diagonal

$$-\eta^{11} = \eta^{22} = \eta^{33} = -1 \tag{11}$$

$$\eta^{jk} = 0; \quad j \neq k.$$

With this understanding, the product $\bar{\Psi}\gamma_5\Psi$ is *well-defined* mathematically and it is plausible that the perturbation expansion makes sense term by term. We now define the free Lagrangian and the interaction term as follows:

$$\mathscr{L} = \sum_j \bar{\psi}^{(j)}(i\slashed{\partial} - m)\psi^{(j)} + \tfrac{1}{2}(\partial^\nu\phi)(\partial_\nu\phi) - \tfrac{1}{2}\mu^2\phi^2.$$

$$= gc^{(j)}c^{(k)}\bar{\psi}^{(j)}\gamma_5\psi^{(k)}\phi + (m^{jk} - m_0^{(j)}\delta^{jk})\bar{\psi}^{(j)}\psi^{(k)}. \tag{12}$$

The mass renormalization is to be carried out in essentially the same form as usual (except that the counter terms are matrices) and we could proceed to calculate the 'large' S-matrix by perturbation theory in the interaction representation:

$$S = 1 + \sum_{n=1}^{\infty} \frac{i^n}{n!} \int d^4x_1 \cdots \int d^4x_n T(\mathscr{H}(x_1) \cdots \mathscr{H}(x_n)). \qquad (13)$$

The Feynman rules for this theory can be deduced from this expansion. We see that from every normal fermion line in a Feynman diagram we have the same diagram with a fermion line corresponding to each auxiliary field in place of it. In particular, since the internal lines come from the 'contraction' of two fermion fields if we absorb the factors $|c^{(j)}|^2$ into the internal lines from the fields the vertex can be considered independent of j. We can thus deal with one diagram with an effective propagator (which is also the contraction function for the effective field Ψ'):

$$S(x - y) = |c_1|^2 \, S(x - y; m_0^{(1)}) - |c_2|^2 \, S(x - y; m_0^{(2)})$$
$$- |c_3|^2 \, S(x - y; m_0^{(3)}) \qquad (14)$$

for each internal fermion line to represent all diagrams with three different kinds of internal lines. If we write the ordinary fermion propagator (for mass m) in momentum space in the form:

$$S(\not{p}; m) = (\not{p} - m + i\epsilon)^{-1}$$

the effective propagator becomes:

$$S(\not{p}) = |c_1|^2 \, (\not{p} - m_0^{(1)} + i\epsilon)^{-1} - |c_2|^2 \, (\not{p} - m_0^{(2)} + i\epsilon)^{-1}$$
$$- |c_3|^2 \, (\not{p} - m_0^{(3)} + i\epsilon)^{-1}. \qquad (15)$$

In order to remove ultraviolet divergences it is sufficient to require that the effective propagator falls off as fast as p^{-3} for large values of the momentum. This can be achieved by the two regularization conditions:

$$\sum_{j,k} \eta^{jk} c_j c_k = 0; \qquad \sum_{j,k,l} \eta^{jk} m_{kl} c_j c_l = 0. \qquad (16)$$

As long as the masses are different these equations can be solved for the c, to obtain:

$$[c^{(2)}/c^{(1)}]^2 = \frac{m_0^{(1)^2} - m_0^{(3)^2}}{m_0^{(2)^2} - m_0^{(3)^2}}; \qquad [c^{(3)}/c^{(1)}]^2 = \frac{m_0^{(1)^2} - m_0^{(2)^2}}{m_0^{(3)^2} - m_0^{(2)^2}}.$$

and the effective fermion propagator becomes:

$$S(p) = |c_1|^2 \, (m_0^{(1)^2} - m_0^{(2)^2})(m_0^{(1)^2} - m_0^{(3)^2}) \prod_{j=1}^{3} (\not{p} - m_0^{(j)} + i\epsilon)^{-1}. \qquad (17)$$

With this propagator all integrals (except the uninteresting vacuum loops) over internal momenta converge.

It is interesting to observe that the fast decrease of the propagator can be thought of as a *superconvergence* requirement so that the regularization conditions are the corresponding *sum rules*. Similar superconvergence requirements and sum rules are expected for the complete propagator in this theory.

A treatment of this kind for the internal lines automatically leads to the 'negative norm particles' (i.e. quanta of the fields with the 'wrong' sign of the commutation relations) in the external lines. These *'ghost' particles* in the initial and final states are dictated by the Lagrangian theory; they are *essential* for preserving the unitarity of the 'large' S-matrix with respect to the indefinite metric. However their inclusions in the initial and final states of a physical process leads to a variety of problems with the probability interpretation. We must therefore deal with the question of the selection of physical states.

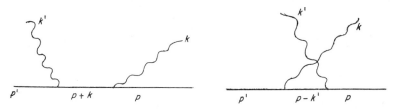

Boson–fermion scattering diagrams.

Let us consider, for illustration, the elastic boson–fermion scattering. In the lowest order, two diagrams contribute to this process. They yield a scattering amplitude which may be written (apart from a numerical factor) in the form

$$M = \bar{u}(p')\gamma_5\{S(p + k) + S(p - k')\}\gamma_5 u(p). \qquad (18)$$

If we choose the fermion and boson masses such that the higher mass fermion states cannot be produced at the energy for the collision, the external lines correspond to the physical fermion only. The S-matrix element would therefore be not only unitary with respect to the indefinite metric but with respect to the positive definite metric of the physical states also. This amplitude has poles in the energy squared variable $s = (p + k)^2$ at the values corresponding to the squares of each of the masses. We note that this scattering amplitude is crossing symmetric.

As the energy of the collision is increased it becomes energetically possible to produce the ghost particles. Even in the second order of perturbation we will have S-matrix elements that correspond to the reactions:

normal fermion + boson $\rightarrow$ normal fermion + boson
normal fermion + boson $\rightleftarrows$ ghost fermion + boson
ghost fermion + boson $\rightarrow$ ghost fermion + boson

The large S-matrix would be isometric with respect to the indefinite metric but would not be probability-conserving for the normal-to-normal transition alone. To obtain a consistent theory we select the *eigenstates of the large S-matrix with positive norm*[2]. These states will form a subspace which has a positive definite scalar product (by construction!). Hence the S-matrix restricted to this subspace will have to be unitary and hence probability-conserving. Let us carry out this calculation also to second order. In this order the state

$$\{b_1\psi^{(1)}(p) + b_2\psi^{(2)}(q') + b_3\psi^{(3)}(q'')\} |0\rangle$$

is an eigenstate of the S-matrix provided q' and q'' are chosen to yield

$$(q' + k)^2 = (q'' + k)^2 = (p + k)^2 = s$$

and b_1, b_2, b_3 are three constants to be determined. There would be three such eigenstates which would reduce to the three fermions as the coupling constant approaches zero. The detailed calculations are somewhat lengthy to be reproduced here but it involves nothing except diagonalization of a 3×3 matrix. The determination of the stationary states of the S-matrix is not altered in principle by proceeding to a higher order in the perturbation expansion[14]. The large S-matrix element is a 3×3 matrix in the fermion type index j. We diagonalize this matrix and select that state which reduces to the physical (positive norm) fermion channel when the coupling constant is made to approach zero.

It is in this identification of physical states that this theoretical framework differs essentially from regularization theory. And it is a consequence of our insistence that the regularization of propagators be due to the extension of the linear space to contain an indefinite metric. In a Lagrangian scheme the modification of internal lines must be inevitably associated with a modification of the treatment of the external lines.

These considerations apply equally well to other theories. If we consider only self-interacting bosons, only one ghost field needs to be introduced for each normal field to make the theory meaningful.

Particle notions in an interacting field theory

It is generally assumed that physically interesting relativistic quantum field theories are capable of a particle interpretation and that all dynamical statements can be transcribed in terms of particle interactions and scattering amplitudes. So far no one seems to have succeeded in constructing an interacting relativistic field theory free from mathematical objections. If it becomes necessary to proceed to an indefinite metric theory to eliminate the mathematical difficulties, we should reexamine the particle notions in field theory to be able to see to what extent the physical identification discussed above is compatible with them.

The vacuum state is invariant under all Lorentz transformations and the one-particle states (belonging to a definite particle type) constitute irreducible manifolds with respect to the Poincaré group. These states are 'steady' and nondegenerate: that is, a state with a definite value of the energy, helicity and linear momentum (together with any other 'internal' symmetry label that may be relevant) belongs to only one such manifold. The multiparticle states are not irreducible; they are not steady and their 'composition' in terms of the constituent particles may change by virtue of scattering.

It is easy to specify what is meant by a two-particle system when they do not interact. For an interacting system we may say that a certain state belongs to a two-particle system if there exists a manifold of states (of the interacting system) containing the state in question and closed under all relativistic transformations which is isomorphic to a manifold of states (within the same values of the momentum and helicity) of a noninteracting system of two particles. The need to introduce such an elaborate definition is that in a quantum field theory the states are not defined directly in terms of particle observables. So that particle concepts have to be introduced from outside and are to a certain extent arbitrary. The *particle interpretation* of a field theory is not uniquely defined by the field theory alone but *depends* on the choice of the construction of *particle variables*.

Let us recall the configurational notions associated with a two-particle state. The 'distorted two-particle wavefunction' is the expansion coefficient for the state of two-particle system in terms of the two-particle states of the noninteracting comparison system. There is also the intuitive notion of measurement of one-particle properties 'when the other particle is far away'. This implies the existence of classical apparatus which converts the two-particle system into a one-particle system; the measurement of one-particle properties can then be made on the state so prepared. The detection of *one-particle properties* thus *presupposes an operation* by means of which the other particle may no longer belong to the quantum mechanical state on which the one-particle measurements are performed.

Within the indefinite metric framework some of these characterizations have to be modified. The no-particle, one-particle and multi-particle states can still be identified, though the representations of the Poincaré group are not necessarily unitary. But those states that can be interpreted as physical states must furnish unitary representations; the particle interpretation must be in accordance with this limitation. We must also satisfy the intuitive notion that a physical one-particle state can be produced by 'removing the other particle' from a two-particle state. Now, the physical two-particle state was identified by taking 'steady states' (i.e. eigenstates of the S-matrix) with positive norm. Since, by definition, physical operations can connect physical states with physical states only, the operation of isolating a component ghost particle cannot be physical. *This nonanalyzability* of the physical states *is fundamental* to the framework and distinguishes these states from the analyzable eigen-amplitudes for scattering of two coupled physical channels.

The treatment of the external lines thus developed is the following[2]: Compute all relevant transition amplitudes in the theory and thus construct the generalized S-matrix of the theory to any desired degree of approximation. No infinities are encountered at any stage of the perturbation series calculation. The physical states are now identified with the positive norm eigenstate of the large S-matrix which reduces to the two-particle state of positive norm particles in the limit of vanishing coupling.

Relation of indefinite metric theory to nonlocal theories

We can observe that the effect of introducing the indefinite metric and the ghost particles is to change the effective fermion propagator and simulates a nonlocal coupling. Consider for example the equation of motion for the fermion fields

$$(i\partial\!\!\!/ - m_j)\psi^{(j)}(x) = \sum_k g\bar{c}_j c_k \phi(x)\gamma_5 \psi^{(k)}(x) = g\bar{c}_j \phi(x)\gamma_5 \xi(x) \qquad (19)$$

where

$$\xi(x) = \sum_k c_k \psi^{(k)}(x)$$

$$\bar{c}_j = c_j \eta_j = \begin{cases} +c_j\,; j = 1 \\ -c_j\,; j = 2, 3. \end{cases}$$

These equations have the (formal) solution:

$$\psi^{(j)}(x) = \psi^{0(j)}(x) + g\bar{c}_j \int S_j(x - y; m_j)\gamma_5 \phi(y)\xi(y)\, \mathrm{d}^4 y. \qquad (20)$$

This could be transcribed into the form

$$\xi(x) = \xi^0(x) + g \int G(x, y)\gamma_5\phi(y)\xi(y)\, d^4y$$

$$\zeta_1(x) = \zeta_1^0(x) + g \int H_1(x, y)\gamma_5\phi(y)\xi(y)\, d^4y$$

$$\zeta_2(x) = \zeta_2^0(x) + g \int H_2(x, y)\gamma_5\phi(y)\xi(y)\, d^4y$$

where

$$G(x, y) = \sum_j \eta_j S_j(x - y)$$

$$H_1(x, y) = \sum_j a_j \eta_j c_j S_j(x - y)$$

$$H_2(x, y) = \sum_j b_j \eta_j c_j S_j(x - y)$$

$$\zeta_1(x) = \sum_j a_j \psi^{(j)}(x)$$

$$\zeta_2(x) = \sum_j b_j \psi^{(j)}(x).$$

By a suitable choice of the parameters c_j we make $G(x, y)$ have a Fourier transform that falls off as fast as p^{-3}. The quantities $H_1(x, y)$ $H_2(x, y)$ do not fall off as fast; at least one of them can fall off no faster than p^{-1}. But the only essential propagation function is $G(x, y)$. This regularized $G(x, y)$ is related to the ordinary Green's function $S(x - y, m_1)$ by an integral transform:

$$G(x, y) = \int K(x, z) S((z - y), m_1)\, d^4z \tag{21}$$

with

$$K(x, z) = (2\pi)^{-4} |c_1|^2 \int \frac{(m_1 - m_2)(m_1 - m_3)}{(\not{p} - m_2)(\not{p} - m_3)} e^{ip(x-z)}\, d^4p. \tag{22}$$

The kernel $K(x, z)$ may now be seen to be translation and Lorentz invariant, and has the effect of eliminating the light cone singularities of $S(x, y)$ of the δ and δ' types to yield the regularized function $G(x, y)$ with no singularities on the light cone[13]. This is the simplest way to see that the indefinite metric theory is essentially a relativistic nonlocal theory.

In spite of this effective nonlocal interaction the analytic structure of the amplitude and the dispersion relations satisfied by the scattering amplitudes can be exhibited. We have already seen that, at least in perturbation theory, the amplitudes have the usual form apart from a regularization of the propagators. Hence (to the extent that the renormalized masses continue to be real) the analyticity domain derived for perturbation theory amplitudes would be valid for the indefinite metric theory also. Once the physical states have been identified as being spanned by the steady states with positive norm, we can get the usual unitarity requirement on these amplitudes. In particular, if the theory is time-reversal invariant we could relate the imaginary part of the amplitude to quantities bilinear in the amplitude[15]. Thus 'causality' (i.e. upper half-plane analyticity) *would be valid* for a time-reversal invariant indefinite metric theory even though it has an effective nonlocal interaction!

It can in fact be shown in simple models that the solutions for scattering amplitudes are *analytic* functions of the square of the coupling constant and can be analytically continued to obtain the solutions for an indefinite metric theory[15]. Once we have obtained an amplitude in the indefinite metric theory it is necessary to make a proper identification of the physical states and this will introduce new analytic functions of the *external* momenta as factors in the definition of the physical scattering amplitude. It appears that these constructions can be implemented in a realistic theory of interacting fields[14].

The relationship between the local indefinite metric formalism and the nonlocal positive metric formalism can be illustrated by a parallel situation in (first quantized) relativistic theory. For describing a spin $\frac{1}{2}$ particle of finite mass in relativistic theory we need a two-component wave function. For an interacting spin $\frac{1}{2}$ particle, say the electron in an (external) electromagnetic field we could still make use of the same scheme, at least when the external field is not too strong. However the description of neither the free particle nor of the interacting system can be manifestly covariant. And relativistic invariance demands that the electromagnetic interaction be nonlocal and nonlinear. This was known from the work of Darwin[17] and seen most clearly from the Foldy–Wouthuysen–Tani transformation[18]. On the other hand the increase in the number of components from two to four enables Dirac to write down a covariant wavefunction; and the electromagnetic interaction is linear, local and manifestly covariant. But the price we have to pay is to have a large space, not all vectors of which represent physical states.

It may be argued at this point that the relativistic first quantized Dirac equation is not the correct description since, for one thing, it breaks down for sufficiently strong external fields: and, for another, it does not describe the second quantized aspects of the electron. But this shows further why the covariant form in the larger space is to be preferred: it is in terms of this form that the further theoretical developments are most satisfactorily formulated.

The parallel is even closer when we realize that the identification of the two-component (positive energy) particle wavefunction in terms of the Dirac wavefunction is made by solving the dynamical equations in the covariant form either exactly or to a suitable degree of approximation. The physical particle wavefunctions for the interacting particle always contains an admixture of the 'negative energy' wavefunction. But the physical particle is identified with this linear combination: the 'negative energy' components are nonanalyzable essential parts of the physical state. This is similar to the situation with indefinite metric theories where the physical states contain a certain 'admixture of the negative norm states' determined by the strength of the interaction. We also find that the nonlocality and nonlinearity of the interaction of the electron with the electromagnetic field should not contain an arbitrary form factor; the formfactor must be *derived* from an *extended* wavefunction *with linear local* coupling. The magnitude of the anomalous magnetic moment and the Darwin term, for example are uniquely determined from the mass and charge. The (three-dimensional) 'form factor' is dynamically derived. The 'form factors' of an indefinite metric theory are also dynamically derived and parallel the form factor of the two-component electron.

Heisenberg's theory

In the theory of elementary particles developed by Heisenberg and collaborators[13] the problem of regularizing the two-point propagation functions has to be tackled. Heisenberg makes use of a regularization prescription in which there are two states of opposite norm which become degenerate in mass (or energy). In this case we get a 'dipole ghost' state. Out of the two states only one is truly an eigenstate of the Hamiltonian (or mass operator!). The other state is a 'dipole'. Out of the various possible solutions of the Schrödinger equation corresponding to scattering states, Heisenberg *selects* out those *states* which correspond to *no asymptotic contribution of the dipole ghost* states. Heisenberg has shown[16] how this programme can be carried out in a suitable version of the Lee model. And it is very plausible that the same method will be applicable for general field theory amplitudes provided a dipole ghost exists in the renormalized mass spectrum. In view of its possible relevance to the photon, in his theory of elementary particles Heisenberg chooses the dipole to be at zero mass[13].

We note several points of similarity of this scheme with the one outlined above. In both of them the field theory finds a simple expression in terms of the extended indefinite metric space. It is in this space that we have local linear operators and local interactions. The physical states are identified, in each approximation, after the amplitudes in the large space are calculated. The physical states are then to be specified after the interaction is specified. In the model that we have discussed the ghost masses are large and hence yield effective couplings which depart by a small amount from locality, while putting the dipole ghost at zero mass makes the departure of the effective coupling from locality somewhat more pronounced. There is, of course the basic difference in the aim of Heisenberg's theory from usual field theory in that it attempts to deduce the physics of elementary particles from a single equation for a single field.

Regularization theory and its field-theoretic formulation

Since there are some similarities between the indefinite metric theory and the process of regularization[11] according to Pauli and Villars and Feynman we give a brief outline of regularization theory and its field theoretic formulation and we contrast it with the indefinite metric theory.

As originally formulated, regularization is a process carried out for S-matrix elements[11], or more generally for the expectation values of operators.[19] Let us consider an amplitude with several external meson lines with momenta $k_1, k_2 \ldots$, and several external nucleon lines with momenta $p_1, p_2 \ldots$, and a suitable number of internal lines. Covariant perturbation theory would then lead to an expression for the contributions of this diagram which would be a function of the external momenta and the masses μ, m of the meson and nucleon respectively and would involve a suitable number of momentum integrations, which may or may not be divergent. Let us denote this expression by $F(k, p; \mu, m)$. Then the regularized expression for this quantity is given by[11]

$$F_R(k, p) = \int d\lambda \int d\kappa \rho(\lambda, \kappa) F(k, p; \lambda, \kappa) \qquad (23)$$

where $\rho(\lambda, k)$ is a real quantity which satisfies the property

$$\int d\lambda \int d\kappa \rho(\lambda, \kappa) = 0 \qquad (24)$$

together with such additional properties that make the quantity $F_R(k, p)$ free of infinities. It is hoped that the ambiguity in the choice of $\rho(\lambda, \kappa)$ would only affect the charge and mass renormalization constants. In such a case the limit of the theory may be taken in which $\rho(\lambda, \kappa)$ consists of a delta function $\delta(\lambda - \mu)\delta(\kappa - m)$ plus contributions from only infinitely large values of λ and/or κ. The renormalized perturbation series is formally unaltered, but the regularized theory works at all stages only with finite quantities.

Several remarks are in order. First of all, regularization is a prescription rather than a modification of the basic theory and as such does not remove the basic difficulties of the field theory. Rohrlich[19] has shown, for example, that the difficulties with the self-stress of the electron are not removed by the usual regularization procedure. Second, the requirement that $\rho(\lambda, \kappa)$ be taken such that $F_R(k, p)$ be free of infinities *cannot* be satisfied with a single choice of $\rho(\lambda, \kappa)$ if $F(k, p)$ contains contributions of all different orders in the coupling constant that are allowed. This happens because the divergent part of the matrix element $F(k, p)$ is itself a power series in the coupling constant, and the $\rho(\lambda, \kappa)$ that leads to finite results in one order will not in general lead to finite results in the next order. This difficulty was not explicitly encountered in the original papers since only second order contributions were explicitly computed!

To illustrate the regularization prescription we compute the second order nucleon and meson self energies for the local field-theoretic interaction (5) for the system described by equations (4)–(7). Apart from inessential factors the nucleon self energy to second order is given by

$$K(p; m) = g^2 \int \gamma_5 (\not{p} - \not{q} - m)^{-1} \gamma_5 (q^2 - \mu^2)^{-1} \, d^4q \tag{25}$$

which diverges logarithmically. We now choose a regulator weight

$$\rho(\lambda, \kappa) = \delta(\lambda - \mu)\{\delta(\kappa - m) + \Sigma a_j \delta(\kappa - m_j)\} \tag{26}$$

$$\Sigma a_j = -1$$

so as to eliminate this divergence. The effect of introducing the regulator weight (26) is to replace the nucleon propagator by a modified propagator:

$$\tilde{S}(x - y) = \tilde{S}(x - y; m) + \Sigma a_j \tilde{S}(x - y, m_j)$$

$$\Sigma a_j = -1 \tag{27}$$

This modification coincides with the indefinite metric theory result with

$$a_j = |c_j|^2 \eta^{jj} \tag{28}$$

But this is the only case where the regularization prescription coincides with the indefinite metric theory! The meson self-energy is proportional to the expression

$$\Lambda(k; m) = g^2 \int \mathrm{Tr}\, \{(\not{q} - \not{p} - m)^{-1} \gamma_5 (\not{p} - m)^{-1} \gamma_5\} \, d^4p \tag{29}$$

which is quadratically divergent. The regularization prescription replaces this by the quantity

$$\Lambda_R(k) = \Lambda(k; m) + \sum_j a_j \Lambda(k, m_j)$$

$$= g^2 \int \mathrm{Tr}\left\{\sum_{j,k} a_j(\slashed{q} - \slashed{p} - m_j)^{-1}\gamma_5(\slashed{p} - m_j)^{-1}\gamma_5 \right.$$

$$\left. + (\slashed{q} - \slashed{p} - m)^{-1}\gamma_5(\slashed{p} - m)^{-1}\gamma_5\right\} \mathrm{d}^4 p. \qquad (30)$$

On the other hand, the indefinite metric theory would yield

$$g^2 \int \mathrm{Tr}\left\{\sum_{j,k} \bar{a}_j a_k (\slashed{q} - \slashed{p} - m_j)^{-1}\gamma_5(\slashed{p} - m_k)^{-1}\gamma_5\right\} \mathrm{d}^4 p. \qquad (31)$$

This already shows that the indefinite metric theory yields a different method of making the self-energy (and more generally transition amplitudes) finite. We also note that since in second order the coupling constant factorizes out the conditions on the auxiliary masses m_j for regularization are *independent of the coupling constant*. This will no longer be true if we considered the self energy including second and fourth order terms, for example. If we have to regularize the amplitude to all orders the restrictions on the auxiliary masses will involve explicitly infinite series in the coupling constants. This is to be contrasted with the indefinite metric theory where the superconvergence requirements (16) and (17) are independent of the coupling constants and eliminate the divergence in every order.

While the usual Lagrangian formalism is not applicable[20] to the regularization prescription, it is possible to construct a quantum field theory in which the Fourier transforms of the vacuum expectation value of the time-ordered products yields the regularized amplitudes. The field theory so constructed has a degenerate vacuum and does not satisfy the usual cluster decomposition property; consequently the scattering amplitude computed from such a theory is not likely to be unitary. Nor are the field operators a complete set of operators. All these again point to the essential difference of the regularized theory from the indefinite metric theory.

Let $\psi_{(x)}^{(\nu)}(x)$, $\phi_{(x)}^{(\nu)}(x)$ be a set of fermion and boson fields which for each index ν satisfy the usual canonical anticommutation and commutation relations:

$$\begin{aligned}
\delta(x^0 - y^0)\{\psi_r^{(\nu)\dagger}(x), \psi_s^{(\nu)}(y)\} &= \delta_{rs}\delta(x - y) \\
\delta(x^0 - y^0)[\phi^{(\nu)}(x), \partial^0 \phi^{(\nu)}(y)] &= i\delta(x - y) \\
\delta(x^0 - y^0)[\phi^{(\nu)}(x), \psi^{(\nu)}(y)] &= 0 \\
\delta(x^0 - y^0)[\partial^0 \phi^{(\nu)}(x), \psi^{(\nu)}(y)] &= 0.
\end{aligned} \qquad (32)$$

The fields for different values of ν commute at *all times*. For each value of ν we consider a Hilbert space $H^{(\nu)}$ in which the fields $\psi^{(\nu)}(x)$, $\phi^{(\nu)}(x)$ are linear operator-valued distributions. We form the space H which is the direct sum of the spaces $H^{(\nu)}$ and define the projection operators $\mathscr{P}^{(\nu)}$ such that

$$H = H^{(1)} \oplus H^{(2)} \oplus \cdots . \qquad (33)$$
$$\mathscr{P}^{(\nu)} H = H^{(\nu)}$$

The elements **X** of H are therefore collections of vectors $X^{(1)}$, $X^{(2)} \cdots$ in the spaces $H^{(1)}$, $H^{(2)} \cdots$. We can make H into an inner product space by the definition

$$(\mathbf{X}, \mathbf{X}) = \sum_\nu [\eta^{(\nu)}/|\eta^{(\nu)}|] \cdot (X^{(\nu)}, X^{(\nu)}) \qquad (34)$$

where $\eta^{(\nu)}$ is a set of real non-zero numbers. Unless all the $\eta^{(\nu)}$ are positive H is an indefinite metric space.

We now introduce the fields $\Psi(x)$ and $\Phi(x)$ in the space H by the relation

$$\mathscr{P}^{(\nu)} \Psi(x) \mathscr{P}^{(\nu')} = \psi^{(\nu)}(x) \mathscr{P}^{(\nu)} \delta_{\nu\nu'} \qquad (35)$$
$$\mathscr{P}^{(\nu)} \Phi(x) \mathscr{P}^{(\nu')} = \phi^{(\nu)}(x) \mathscr{P}^{(\nu)} \delta_{\nu\nu'}$$

If we introduce the vacuum state Ω by the relation

$$\mathscr{P}^{(\nu)} \Omega = c_\nu \Omega^{(\nu)} \qquad (36)$$

where $\Omega^{(\nu)}$ is the vacuum state in $H^{(\nu)}$, then it follows that the regularized vacuum expectation value of any product of field operators $\psi, \bar\psi, \phi$ is obtained by taking the expectation value of the corresponding product of the operators $\Psi, \bar\Psi, \Phi$ in the state Ω. In particular we consider the interaction density

$$V(x) = g\bar\Psi(x)\gamma_5\Psi(x)\Phi(x) \qquad (5^1)$$

the quantities

$$S_n = i^n/n! \int d^4x_1 \cdots \int d^4x_n (\Omega, T(V(x_1) \cdots V(x_n))\Omega)$$

give the regularized S-matrix element to the nth order in the perturbation, where the regularization weight function $\rho(\lambda, \kappa)$ is given by

$$\rho(\lambda, \kappa) = \sum_\nu \eta_\nu \delta(\lambda - \lambda^{(\nu)}) \delta(\kappa - \kappa^{(\nu)}).$$
$$\eta_\nu = |c_\nu|^2 \cdot (\eta^{(\nu)}/|\eta^{(\nu)}|) \qquad (37)$$

The regularization can be applied either to S-matrix elements or to expectation values of products of field operators.

This construction is a simple extension of the method of convex sums[21] and much of the structure of the convex model field theories are reproduced here. In particular the cluster decomposition property no longer obtains since the vectors Ω' with

$$\mathscr{P}^{(\nu)}\Omega' = \pm\Omega^{(\nu)}$$

are also equally satisfactory as vacuum states. Any one of these vectors is 'cyclic', i.e. the operator $\Psi'(x), \overline{\Psi}'(x), \Phi(x)$ acting on Ω' can produce all the vectors in H. But the fields do not constitute a complete set of operators since the vector $\Omega + \Omega'$ for example is *not* cyclic.

General field-theoretic questions

We saw that regularization by indefinite metric which eliminates ultraviolet divergences at the same time makes the product of field operators at the same point mathematically well-defined. It is interesting to enquire about general field theoretic questions within such a framework. We have already seen that the time ordered products are covariant, and that no ambiguities are inherent in the use of local interactions. The theory is therefore manifestly Lorentz-invariant. According to standard methods of quantization (extended to include the indefinite metric) the fields obey local commutation (or anticommutation) relations.

Some time ago Wightman showed that the vacuum expectation values of products of field operators uniquely define the field theory[19]; it is possible to reconstruct the field operators using the method of Segal and Gelfand, and that the manifest Lorentz invariance of the theory was fully equivalent to the manifest Lorentz invariance of the vacuum expectation values. It is easy to see that exactly the same situation obtains for the algebraic structure in a theory with indefinite metric. The vacuum expectation values are manifestly covariant functions which uniquely define the field operators. This correspondence is however only with regards to the algebraic structure. The analytic structure would be the same (future-tube analyticity) only if complex eigenvalues for energy and momentum are excluded by an explicit spectral postulate. Of course in this case the fields are not operators (or operator-valued distributions) in a Hilbert space but they act in a more general inner-product space. (The notion of Cauchy sequences and limits in such a space has to be different from that for a Hilbert space!).

It has been found that within the framework of a positive definite metric the commutator of the various components of a conserved (or 'partially conserved') operator may not vanish as deduced from a formal calculation. This fact has been noted by many people like Goto and Imamura, Pradhan, Källen[23]. Schwinger[24] has given a very simple demonstration of how this comes about for the commutator of the charge density with the current density: Let us assume

$$\delta(x^0 - y^0)[\rho(x), \mathbf{j}(y)] = \mathbf{c}(x, y). \tag{38}$$

Then, by virtue of the continuity equation

$$\dot{\rho} + \nabla \cdot \mathbf{j} = 0 \tag{39}$$

we could deduce:

$$\delta(x^0 - y^0)[\rho(x), \dot{\rho}(y)] = -\nabla \cdot \mathbf{c}(x, y).$$

Taking vacuum expectation values of both sides and recalling that the vacuum is a zero energy state:

$$\delta(x^0 - y^0) \langle 0| \rho(x) \mathcal{H} \rho(y) |0\rangle = - \langle 0| \nabla \cdot \mathbf{c}(x, y) |0\rangle. \tag{40}$$

The left hand side cannot vanish and hence the commutator cannot vanish, since the Hamiltonian H is positive definite and the expectation value on the left hand side is positive definite. This demonstration can be extended to a variety of other cases to demonstrate the need for 'Schwinger terms'.

We now point out that this situation no longer obtains in a properly regularized indefinite metric theory. The above demonstration fails because the left hand side expression could vanish even with H having only nonnegative eigenvalues. A direct calculation of course shows that these charge and current densities (for spinor fields) are well defined and commute with each other.

We must point out that though the general vacuum expectation values of (ordinary and time-ordered) products of field operators (are local and covariant and hence) are crossing symmetric, this is no longer true for the physical scattering amplitude. This comes about because the diagonalization of the S-matrix in the large space to select out the physical states is *not crossing symmetric*. We have thus additional kinematic factors modifying the field-theoretic crossing property. Of course these diagonalizations are irrelevant below the threshold for the production of ghosts, but they have to be taken into account for a general amplitude. Needless to say, such a possible kinematic factor cannot be put to direct experimental test at the present time.

Concluding remarks

The use of the indefinite metric removes the divergence difficulties of field theories and removes mathematical ambiguities in the definition of the equations of motion. This makes it possible to consider a perturbation expansion of the predictions of the theory. The question of convergence of such a power series expansion is not obvious, though there is no obvious reason why it should not converge. But apart from the question of convergence is the more practical question of the goodness of a low order perturbation theory approximation. But this clearly depends on the particular field theory under consideration.

For a conventional treatment of elementary particle interactions it is necessary to exercise considerable judgement in the choice of the primary fields and the primary interactions. Present ideas in particle physics distinguish three categories of interactions, strong, electromagnetic and weak. We have shown elsewhere that by classifying the fields into baryon fields, vector and axialvector meson fields, electromagnetic field and lepton fields we could formulate a simple theory of primary particle interactions in which a single coupling constant characterizes each category of interaction[25]. The dynamical predictions of such a theory would require explicit Lagrangian perturbation theory computations. It has also been found possible to devise a finite quantum electrodynamics of electrons based on an indefinite metric[14]. In a more ambitious theory one may hope to be able to calculate the mass levels and even derive the various categories of interaction as a consequence of a single fundamental interaction structure. In any case, it appears that a finite relativistic quantum theory of fields could be constructed in a simple fashion and without mathematical ambiguities within the framework of an indefinite metric space. The use of an indefinite metric may be viewed as a method of restricting the *number* of degrees of freedom of a field. The free quantized field, as well as the free classical field, has an infinite number of degrees of freedom. The indefinite metric formulation that we have discussed above has the virtue of making the number of effective degrees of freedom (for the fermion field) finite. It is curious to remember that the application of the notions of statistical mechanics to the *free* radiation field in classical theory led to manifest absurdities like infinite specific heats and energy densities at any temperature, and led one to expect a constant specific heat independent of temperature. To reconcile ourselves with nature, we had to search for a theoretical formulation in which not only was the specific heat finite but dependent on the temperature. *When classical theory was replaced by quantum theory the effective number of degrees of freedom at any temperature became finite and the divergence difficulty connected with the specific heat was eliminated.* Thus the *divergence was intimately connected with the infinite number of effective degrees of freedom* for the free field.

We now find that where interacting fields are concerned we have *divergences in the quantum theory*, and that they *must be eliminated by a further restriction of the effective quantum degrees of freedom*. The indefinite metric that we have introduced does precisely this: the effective quantum degrees of freedom are made finite and this means that the divergences of the former theory are eliminated. There are potentially infinite number of degrees of freedom but they must be dynamically constrained to be finite. The use of local field theory automatically leads to the potential infinity of the number of degrees of freedom; and the indefinite metric seems to be a natural way of constraining them.

This line of thinking would suggest that *the indefinite metric formulation of quantum field theory is the prelude to a new theory* in much the same way as the quantum hypothesis was the prelude to quantum mechanics. This suggestion is strengthened by the observation that one possible way of viewing quantum theory is in terms of (commuting) phase space variables, but in such a case the phase space densities are in general indefinite: they take positive and negative values. The equation of motion for the phase space densities are nonlocal and do not in general preserve positive definiteness of the phase space distribution. The lack of positive definiteness of the phase space densities does not lead to negative physical probabilities since the 'physical' measurements are restricted by Heisenberg's uncertainty principle. We have a parallel situation in the indefinite metric quantum field theory: negative mathematical probabilities appear but the particle interpretation of the field theory is so chosen that in physical processes the probability is always non-negative.

The same situation occurs also in connexion with the optical Equivalence Theorem and the Diagonal Representation in quantum optics. In this case the states of the quantized radiation field are displayed in a form which is formally identical with the statistical states of a classical radiation field, but the ensemble density function is not necessarily positive definite.

It is thus likely that the use of the indefinite metric in quantum field theory is a temporary device and that it is a provisional method of discussing a new physical theory. A more satisfactory reformulation would be such that negative probabilities do not appear even in the intermediate steps. But until such a theory is discovered, it seems worthwhile to explore dynamical calculations in an indefinite metric quantum theory.

Acknowledgements

I wish to thank Professor C. F. von Weizäcker and Professor F. Rohrlich for discussions of topics pertaining to this report and to Professor J. R. Klauder and Professor J. Mehra for criticism of the manuscript.

References

(1) Compare the beautiful discussion in P. A. M. Dirac, *Principles of Quantum Mechanics*, Cambridge University Press (1958), Chapter 1.
(2) Sudarshan, E. C. G., *Phys. Rev.*, **123**, 2183 (1961).
(3) In contrast to the mass renormalization, the coupling constant and wave function renormalizations are a matter of convenience.

(4) Heitler's theory is an example of a nonlocal theory which is local-in-time. See Heitler, W., *The Quantum Theory of Fields*, Proc. 12th Solvay Conference, Brussels (1961).
(5) Pryce, M. H. L., *Proc. Roy. Soc. (London)*, **A195**, 62 (1948), Dirac, P. A. M., *Rev. Mod. Phys.*, **21**, 392 (1949) Jordan, Macfarlane and Sudarshan, *Phys. Rev.*, **133B**, 487 (1964).
(6) Wigner, E. P., and Van Dam, H., *Phys. Rev.*, **138**, B1576 (1965).
(7) This result was first deduced for classical theory by Currie, Jordan and Sudarshan, *Rev. Mod. Phys.*, **35**, 350 (1963), and extended by Leutwyler, H.
(8) Hertz, H., *Miscellaneous Papers, Vol. III Principles of Mechanics*, Macmillan, New York, (1896).
(9) Gupta, S. N., *Proc. Phy. Soc.*, **A63**, 681 (1950); **A64**, 850 (1951); Bleuler, K., *Helv. Phys. Acta*, **23**, 567 (1951).
(10) Dirac, P. A. M., *Proc. Roy. Soc.*, **A180**, 1 (1942); Heisenberg, W., *Nucl. Phys.*, **4**, 532 (1951); K. L. Nagy, *Suppl. Nuovo, Cimento*, **17**, 92 (1960); Pandit, L. K., *Suppl. Nuovo Cimento*, **10**, 157 (1959); Schlieder, S., *Z. Naturforsch.*, **15a**, 448 460, 555 (1960); Scheibe, E., *Ann. Acad. Sci. Fennicae, Ser. AI*, **294** (1960).
(11) Pauli, W., and Villars, F., *Rev. Mod. Phys.*, **21**, 434 (1949). Feyman, R. P., *Phys. Rev.* **76**, 749 (1949).
(12) Green, A. E. S., *Phys. Rev.*, **73**, 26 (1948); Pais, A. and Uhlenbeck, G. E., *Phys. Rev.*, **79**, 145 (1950).
(13) Dürr, Heisenberg, Mitter, Schlieder and Yamazaki, *Z. Naturforsch*, **14a**, 441 (1959); W. Heisenberg, *Introduction to the Unified Theory of Elementary Particles*, Wiley, London (1966).
(14) For application to quantum electrodynamics, see: Arons, Han and Sudarshan, *Phys. Rev.*, **137**, B1085 (1965). Four-fermion interaction and the renormalization scheme is treated in: Sudarshan, E. C. G., *Nuovo Cimento*, **21**, 7 (1961).
(15) Schnitzer, H. J., and Sudarshan, E. C. G., *Phys. Rev.*, **123**, 2193 (1961).
(16) Heisenberg, W., *Nucl. Phys.*, **4**, 532 (1957).
(17) Darwin, C. G., *Proc. Roy. Soc.*, **A118**, 654 (1928).
(18) Foldy, L. L., and Wouthuysen, S. A., *Phys. Rev.*, **78**, 29 (1950), Tani, S., *Progr. Theor. Phys.*, **6**, 267 (1951).
(19) Rohrlich, F., *Phys. Rev.* **77**, 357 (1950).
(20) An explicit statement to the contrary is made by S. N. Gupta, *Proc. Phys. Soc.*, **A66**, 129 (1953).
(21) Sudarshan, E. C. G. and Bardakci, K., *J. Math. Phys.* **2**, 767, (1961).
(22) Wightman, A. S., *Phys. Rev.*, **101**, 860 (1956).
(23) Goto, T., and Imamura, Y., *Prog. Theor. Phys.*, **14**, 396 (1955), Pradhan, T., *Nucl. Phys.*, **9**, 124 (1958), Källén, G., Unpublished.
(24) Schwinger, J., *Phys. Rev. Letters*, **3**, 296 (1950).
(25) Sudarshan, E. C. G., '*The Nature of the Primary Interactions of Elementary Particles*', Syracuse University Report NYO-3399-137 (August 1967); Proc. of the Interaction Conference on Particles and Fields, Rochester (1967); (see pp. 53–62 of this volume).

Discussion on the report of E. C. G. Sudarshan

M. Froissart. I would like to inquire about the following aspect of your theory, namely the cluster decomposition property, as it is probably the weakest causality requirement for any theory. Is it true in this theory that the diagonalization of the whole amplitude agrees with the diagonalization of the separate terms of a non-connected amplitude? This is of course ensured in the Gupta–Bleuler formalism, because of the simple algebraic structure of the supplementary condition.

E. C. G. Sudarshan. I agree with your comment about the Gupta–Bleuler formalism: their supplementary condition 'factorizes'. Now, with regard to the cluster decomposition of the amplitude in the indefinite metric theory, it is clear that it has to be true for the 'large S matrix element' since the cluster property depends only on the spectral properties and in particular on the existence of a unique vacuum. It does not directly depend on the metric of the linear space. When you go from the 'large' space to the 'small' space and consider the physical scattering amplitude we might ask if this amplitude has cluster properties. My guess (based on perturbation theory computations) is that it does. But in any case we can state definitely the following: if the 'large' amplitude factorized into the product of two independent amplitudes, the 'small' amplitude will also automatically factorize out into the product of the corresponding two independent amplitudes.

In spite of this, the question for causal factorization (in the sense in which Stückelberg originally introduced it, and which is sometimes referred to as 'one-particle structure' or as 'double scattering structure') has to be investigated anew since the particle interpretation in this theory is somewhat different from that in the usual theory.

H. Umezawa. On the way in which Professor Sudarshan presented his argument on the theory of indefinite metric, it is crucially required that the 'mathematical' fields with negative metric appear, in the small space, only through observable particles, each of which as a whole has a positive probability. It is an important question to ask how this requirement can be satisfied. However, when we assume that this requirement is satisfied, then the cluster decomposition property in the small space can take quite a different form from that in the large space because interpolating fields, which interpolate the particles of positive probabilities in the small space, may be no more the original field operators but be complicate functions of latters. When we talk about the cluster decomposition, it is important to specify the choice of interpolating fields, or in another way of saying, to specify the representation in which observable particles are described. The point is that Sudarshan's choice of representation for observable particles might be the one in which the cluster decomposition property is obtained, not in the whole large space, but in the small space only.

R. Omnès. If one builds a normed wave packet of direct products of two free particles (i.e. of eigenstates of the S-matrix in the one-particle sector) it should be proved that it is a norm 1 combination of normed eigenstates of S in the two-particle sector, otherwise there is no theory of scattering and one cannot describe the states produced by an accelerator.

In other words, relating my question to the one by Froissart, does

$$\langle \alpha'\beta' | S | \alpha\beta \rangle = A\delta_{\alpha\alpha'}\delta_{\beta\beta'} + \langle \alpha'\beta' | T | \alpha\beta \rangle$$

with $A = 1$?

E. C. G. Sudarshan. If one starts with a two-particle state which is normalized and of positive norm (and reduces to the state of two free positive-norm particles in the limit of no interaction) it should preserve its norm on scattering. We should then define the scattering amplitude by

$$S = 1 + iT$$

But in the present theory, the two-particle state is *not* built out of the 'product' of two free particles. It is rather chosen in the following fashion: denote the initial particle-type labels (in the model, this applies only to the fermions) α, β and final particle labels α', β'. We now take the S-matrix element $S_{\alpha\beta,\alpha'\beta'}$ (or the T-matrix element $T_{\alpha\beta,\alpha'\beta'}$) at any required centre-of-mass energy and angular momentum and diagonalize it. This can be done only after the 'large' S-matrix is computed to any desired accuracy by some method, say (mass-renormalized) perturbation theory. Out of these eigen-channels of the 'large' S-matrix we select out those positive metric states which reduce to the state of two free particles with $\alpha = \beta = \alpha' = \beta' = 1$ in the limit of no interaction. This defines the 'small' S-matrix. In the range of energies

$$2m_1 < \sqrt{s} < m_1 + m_2$$

this diagonalization is trivial since only the $\alpha = \beta = \alpha' = \beta' = 1$ matrix elements are non-zero. But as the energy rises and reaches the domain

$$m_1 + m_2 < \sqrt{s} < 2m_2$$

we have the channels $\alpha + \beta = 2$ or $\alpha' + \beta' = 2$ open and we must do a genuine diagonalization.

This has the consequence that the elastic scattering amplitude acquires an additional kinematic factor (and an additional branch point) at the threshold for the unphysical states!

R. Haag. Concerning the question of the cluster property of the S-matrix and locality one should perhaps keep in mind that there is a hierarchy of such properties necessary for macroscopic locality. There is first the cluster property or 'vacuum structure' which corresponds to the distinction of connected vs. disconnected diagrams. This I would expect to be satisfied in this model. The more difficult thing to check and to be satisfied is the single particle structure which demands the correct decomposition of double scattering into single scattering parts.

E. C. G. Sudarshan. The Stückelberg structure which demands the correct decomposition of double scattering into single scattering parts is obtained in the amplitudes *calculated in perturbation theory* for those configurations for which the double scattering becomes a physical process. They have however additional factorizable contributions corresponding to a double scattering structure with a ghost particle connexion between the two scattering parts. This appears inevitable and it is obtained in quantum electrodynamics where we have double scattering structures with longitudinal photon connecting links.

G. Källén. First, I should like to comment that there are more ways to make electrodynamics than the Gupta–Bleuler way and the non local formulation using the Coulomb or radiation gauge. You can construct a completely local and Lorentz invariant formalism by considering electrodynamics as the limit of a theory with a small photon mass (cf. *Handbuch der Physik*, (*Vol. I*), p. 194, Springer 1958).

Next, I have a question: you have made several statements about the behaviour of your formalism. What is the basis for these statements? Are they based on perturbation theory calculation? If so, to which order? If not, what exactly have you calculated?

E. C. G. Sudarshan. In relativistic field theory most of the calculations that I have done are based on a (mass-renormalized) perturbation theory. So far I have carried out the calculations only to fourth order. Arons *et al.* (*Phys. Rev.*, **137B**, 1085 (1965)) have systematically studied quantum electrodynamics within this formalism and shown that the agreement with experiment could be recovered without any infinities. As far the question of interpretation of states and the analytic properties and dispersion relations for the scattering amplitude, Schnitzer and I (*Phys. Rev.* **123**, 2193 (1961)) have studied a number of models without recourse to perturbation theory.

Incidentally, though each term of the perturbation theory result is finite and unambigous in this theory, there is no reason why the infinite series should converge. On the other hand, there is no reason why it should not converge. At the present time the question of convergence of the power series expansion in the coupling constant is an open question.

REPORT ON THE PRESENT SITUATION IN THE NON-LINEAR SPINOR THEORY OF ELEMENTARY PARTICLES

W. Heisenberg

Max-Planck-Institut für Physik und Astrophysik, München, Germany.

The non-linear spinor theory, as an attempt for a more fundamental theory of elementary particles, rests on the conviction that the complicated spectrum of elementary particles—as, for example, in old times the optical spectrum of the iron atom—must finally be deducible from an underlying natural law; and we hope it to be a simple law, whatever its mathematical form may be. From the experimental material available ten years ago, when Pauli and I worked on this problem, it looked as if the spinor equation

$$i\sigma_v \frac{\partial \chi}{\partial x_v} + \sigma_v : \chi(\chi^*\sigma^v\chi) := 0 \qquad (1)$$

(this is the form given to the equation later by Dürr) could possibly be a sufficient frame, a suitable 'master equation' for such a theory. In the meantime, much new information has been collected by the experiments, and the theory has been developed in many details. Therefore a survey of the present situation may be useful.

The first part of my talk will be devoted to the mathematical structure of the theory, the second to the symmetry properties of the equation and their consequences with regard to the experiments; in the third part I will try to compare the methods and results of this theory with those of more conventional schemes.

1. The mathematical scheme

If one wants to give a mathematical meaning to a field equation like (1), the obvious example is quantum electro-dynamics. This latter theory is undoubtedly a working theory, even if its mathematical structure is not completely known; it gives very accurate results, e.g. for the Lamb shift. During the development of this theory in the early thirties we learnt that the postulates of quantum theory and those of special relativity cannot easily be reconciled. The requirement of local causality in special relativity together with the uncertainty relations of quantum theory may cause divergencies, and the long range of the electromagnetic field (rest mass zero of the photon) introduces new problems. A price has to be paid, and it may be paid at different points. First the process of renormalization seems to be an essential part of the formalism. Furthermore one can either introduce an indefinite metric in Hilbert space, as Bleuler and Gupta have done or one can give up the manifest Lorentz covariance of the scheme and introduce the non local Coulomb forces, as in Dirac's theory of radiation; or one may introduce limiting processes as suggested by Källén. All these forms are equivalent. In any case, when the price has been paid, one can construct an approximation scheme, which in every step gives well defined, finite results. It is a characteristic feature of this perturbation theory that in every step the number of variables in the wavefunctions is limited, but it increases indefinitely by going to higher and higher approximations. In principle a hydrogen atom may not only consist of proton and electron, it may also be composed of a proton, 2 electrons and 1 positron, or generally a proton, n electrons and $n-1$ positrons. If all these infinite possibilities were to be included from the beginning, the equations of quantum electro-dynamics would probably not define a mathematical problem. But every step in the approximation scheme does define a mathematical problem, and the complete theory has a meaning, if this approximation scheme converges. A proof of convergence has not yet been given.

Taking this mathematical interpretation of quantum electro-dynamics as a model, one can try to give a mathematical meaning to Eqn. (1) in a similar fashion. If one studies the behaviour of the 2-point function

$$F(x-y) = \langle 0| \chi(x)\chi^*(y) |0\rangle \tag{2}$$

and the rôle played by it in every step of the approximation scheme, one sees that it cannot contain δ- or δ'-functions at the light cone—contrary to the conventional Schwinger functions—otherwise the occurring integrals could diverge. This cancelling of the δ- and δ'-functions can only be achieved by introducing an indefinite metric in Hilbert space. After doing this one may represent (2) by

$$\langle 0| \chi(x)\chi^*(y) |0\rangle = (2\pi)^{-4} \int \rho(\kappa^2)\, d(\kappa^2) \int d^4p\, e^{ip(x-y)} \frac{p_\nu \bar{\sigma}^\nu \cdot \kappa^4}{(p^2)^2(p^2-\kappa^2)}. \tag{3}$$

For a general spectrum $\rho(\kappa^2)$ this representation is actually not less general than the usual representation given by Umezawa, Kamefuchi, Källén, or Lehmann. But for a mass spectrum consisting only of a few lines, or rapidly converging at high masses, it will generally state the existence of a regularizing dipole ghost at mass zero. Therefore in low approximations, when only a few masses can be considered, this representation (3) contains a significant statement concerning the behaviour of the mass spectrum at very small masses. This statement was, as a trial, suggested by the experimental situation and by an argument discussed at this conference by Chew. If one starts from a 'master equation', like (1), there is always the danger that some particles could appear as 'real elementary particles', while from the experiments we have good reason to believe, that 'every elementary particle consists of all other particles', i.e. that actually all particles are compound systems, are 'dressed up' by interacting with the others. In this respect I would agree completely with the philosophy outlined by Chew in the first part of his talk. In case of Eqn. (1) there is obviously the danger that the neutrino could appear as 'really elementary'. Therefore this has at once been excluded by putting a dipole ghost at mass zero, thereby giving the neutrino solution the norm zero. Hence there is in this approximation no real particle of mass zero which could interact with the others. This dipole ghost could therefore, in the lowest approximation, represent the lepton part of the spectrum, which does not take part in the strong interactions; and the hope would be, that in higher approximations the dipole ghost at mass zero would gradually develop into the more complicated real spectrum of leptons, which then interact electromagnetically or weakly with other particles. The consequences of a dipole ghost have been studied in detail with the help of the Lee model.

The actual approximation method can be constructed from different schemes, which have in common the fundamental assumption that, as in quantum electro-dynamics, in every finite approximation the number of variables in the wavefunctions is limited, that this number however increases indefinitely with going to higher and higher approximations. Every single step in the scheme gives well defined, finite results, and if the scheme converges it defines a solution of the problem. The convergence is expected to be much slower than in quantum electro-dynamics since in Eqn. (1) there is no weak perturbation term; again no proof of convergence has yet been given. In the early papers mostly the new Tamm–Dancoff method had been used. Recently considerable help has been obtained from the methods of many body physics, which have proved successful e.g. in the theory of solid bodies. Actually the problems of solid state physics are frequently similar to those of elementary particle physics, since 'excitons' and 'polarons' etc. are also particles 'dressed up' by interaction, and the spontaneous breakdown of symmetries by the ground state may happen equally well in both cases. It is especially the method of the Green's functions, developed by Schwinger and others, which can be used with success in both theories.

NON-LINEAR SPINOR THEORY

I might just mention two examples, one using the Tamm–Dancoff method the other one that of the Green's functions, in order to illustrate the practical applications. The following notation will be used:

$\langle 0| \chi(x)\chi^*(y) |0\rangle =$ —◯—; $\langle 0| \chi(x)\chi(y)\chi^*(z)\chi^*(u) |0\rangle =$ ⊃◯⊂

The irreducible part of ⊃◯⊂ will be called ⊃η⊂.

Wave functions:

$$\langle 0| \chi(x) |\text{Fermion}\rangle = \text{—}\triangleleft \qquad (4)$$

$$\langle 0| \chi(x)\chi^*(y) |\text{Boson}\rangle = \triangleright\triangleleft$$

Green's function of mass zero:

$$(2\pi)^{-4}\int d^4p\, e^{ip(x-y)} \frac{p_\nu \sigma^\nu}{p^2} = \text{——}.$$

The lowest Tamm-Dancoff approximation for the boson eigenvalue equation is

$$\triangleleft = \triangleright\!\circ\!\triangleleft \quad \text{or} \quad \triangleleft = \triangleright\!\circ\!\triangleleft$$

$$\text{or} \quad \left(1 - \circ\!\circ\!\circ\right)\triangleleft = 0 \qquad (5)$$

The lowest approximation for the 4-point function in the method of Green's function is

$$\supset\!\eta\!\subset \;=\; \supset\!\circ\!\eta\!\subset \;+\; \times\!\!\!\circ\!\circ \qquad (6)$$

with the solution

$$\supset\!\eta\!\subset \;=\; \frac{1}{1 - \circ\!\circ} \qquad (7)$$

Equation (5) gives the masses of π- and η-meson in a fair approximation. The low mass of the pion, which had been mentioned as a major problem by Chew, comes out naturally from (5); and this result rests essentially on the dipole ghost at mass zero in (3). Equation (7) shows clearly the poles of the 4-point-function at the masses of the bosons and it gives, according to Dhar and Katayama, a value for the coupling of the bosons to the nucleons in reasonable agreement with the experiments.

The method of the Green's functions allows, at least in principle, the determination of the 2-point function (3) and the spectrum $\rho(\kappa^2)$ from Eqn. (1). Unfortunately already the lowest order approximation is too complicated for an explicit solution. Still one can see from the equation, that there cannot be δ- or δ'-functions at the light cone of (3), and that the asymptotic behaviour defined by the dipole ghost may go well together with the requirements of the equation.

The flexibility of quantum electro-dynamics with respect to the point where the price has to be paid for the reconciliation of quantum theory and relativity suggests the existence of other mathematical schemes interpreting Eqn. (1), which do not introduce an indefinite metric in Hilbert space, but replace it by the concept of non-local forces. A simple scheme of this type has been suggested by Dürr. Instead of the field operator $\chi(x)$ one may introduce another field operator $\psi(x)$ connected with $\chi(x)$ by the relation:

$$\psi(x) = \Box \chi(x) \quad \text{or} \quad \chi(x) = \Box^{-1}\psi(x) + \chi_0(x). \tag{8}$$

For the 2-point function $\langle 0| \, \psi(x)\psi^*(y) \, |0\rangle$ one gets from (3)

$$\langle 0| \, \psi(x)\psi^*(y) \, |0\rangle = (2\pi)^{-4} \int \rho(\kappa^2) \, d(\kappa^2) \int d^4p \, e^{ip(x-y)} \frac{p_\nu \bar{\sigma}^\nu \cdot \kappa^4}{p^2 - \kappa^2}. \tag{9}$$

The dipole ghost at mass zero has disappeared and the spectrum $\rho(\kappa^2)$ could well be positive definite, since now the wave Eqn. (1), written in terms of $\psi(x)$, contains only a non local interaction ($\Box^{-1}$ is a non local operator), which allows the integrals in the approximation scheme to be finite, even if the 2-point function (9) contains δ- or δ'-functions at the light cone. It is true that the non local forces in the wave-equation would give rise to rather complicated problems concerning the boundary conditions, i.e. the in- and out-fields connected with (8), but there may well be a one-to-one correspondence between the solutions written in terms of an indefinite metric and those starting from the non local forces. Therefore the theory established by Eqn. (1) has probably the same kind of flexibility as quantum electro-dynamics.

2. The symmetry properties connected with Eqn. (1)

Equation (1) is invariant under the proper Lorentz group, including the two discrete operations *PC* or *PCT*, the isospin group SU_2, a gauge group which may represent the conservation of the baryonic number and finally the dilatation group. It is not invariant under SU_3 or higher groups of this type, and it does not immediately represent parity P, strangeness (or hypercharge) and lepton conservation. Furthermore the invariance of (1) under the dilatation group cannot lead to an invariance of the complete theory under this group, since an invariant 2-point function or commutator would imply a δ-function in (3), which is impossible. Therefore a mass scale must be introduced; then as a kind of compensation one can define parity, as has been pointed out by Dürr, since for a massive particle one may always define 'left-hand' and 'right-hand' wavefunctions. But in order to represent the empirical spectrum of elementary particles with regard to strangeness and SU_3 it will certainly be necessary to introduce new degrees of freedom.

The most natural way of doing this seems to be the assumption of a spontaneous breakdown of symmetry, the introduction of an asymmetrical groundstate, like in solid state physics; such a breakdown could on the one hand lead to an understanding of the electromagnetic violation of SU_2; on the other hand it would give an isospin property to the vacuum; it would therefore, without changing Eqn. (1), introduce new degrees of freedom from the vacuum, which could be just sufficient to account for hypercharge (or strangeness) and, in a very rough approximation, for SU_3 and the higher groups.

This general scheme has been followed up to some extent in recent years. The most important step was the application of the theorem of Goldstone in a somewhat generalized form (in a paper by Dürr, Yamamoto, Yamazaki and myself). In the special form given to the theorem in the first papers of Goldstone, Salam, Weinberg, Nambu and others, it states that the degeneracy of the vacuum will automatically lead to the existence of particles of mass zero, and thereby indirectly to long range forces. In more recent investigations of Higgs, Kibble and others, which have been reported by Dürr at this conference, the problem has been reversed by asking: if long range forces are assumed from the beginning, what are the consequences of a degeneracy of the vacuum? I need not discuss the answers again. In any case the investigations have revealed an intimate connexion between an asymmetry of the groundstate and long range forces. Therefore in the non linear spinor theory the problem could be formulated by asking: Is it possible that the equations for the Green's functions, which are themselves symmetrical under the isospin group SU_2, could have asymmetrical solutions, and that this asymmetry in isospace could be connected with the appearance of a boson-pole at mass zero, corresponding to the photon in its transformation properties?

This seems in fact to be the case. It turns out that a long range field of this type must be defined with regard to its transformation properties in isospace by a projection operator $\frac{1}{2}(Y + \tau_3)$, in accordance with the rule of Gell-Mann and Nishijima; this projection operator then distinguishes between neutral and charged particles. If the dipole of the charged leptons in (3) is moved from zero to finite masses—thereby indicating the asymmetry in isospace and an electromagnetic mass of the leptons—a boson-pole at mass zero and spin 1 can actually be established from (5) or (7). In that approximation, in which the baryon octet is represented by just one pole in (3), the average lepton mass is determined to be ∼40 MeV, and the coupling, i.e. the fine-structure constant has, according to (7), a value around $\frac{1}{120}$. Therefore it seems that the actual behaviour of nature in this field can be well imitated in the mathematical scheme of the non linear spinor theory.

At the same time the interaction between the particles and the groundstate—which may be studied e.g. by the model of a ferromagnet—can lead to the formation of strange particles i.e. particles, to which some isospin from the groundstate (a 'spurion') has been attached as has been pointed out by Biritz. In this way the hypercharge Y can be established by a gauge transformation in isospace of the groundstate. Finally the two lowest octets in the baryon and the boson spectrum seem to be a natural consequence of the symmetry in the interaction between particles and groundstate.

If this picture is correct, the three main interactions can be described in the following way. In the strong interaction the groundstate never takes up any property of the particles and vice versa; therefore in collision processes pairs of spurions and antispurions with isospin zero may be created or annihilated, but this makes no change in the groundstate. In electromagnetic processes spurion–antispurion pairs of the electromagnetic (photon) type may be transferred between the particles and the groundstate, and these pairs have with equal probability isospin 1 and 0; therefore in such processes isospin is transferred, but no hypercharge. Finally in weak interactions even single spurions (hence isospin and hypercharge) may be transferred. This would look like a process contradicting causality; but it may (according to Dürr) be compared with the Mössbauer effect where a momentum seems to be transferred at once on the whole crystal. In the latter case Weisskopf has demonstrated, that a deviation from causality cannot be observed.

The idea of using the interaction between particles and the groundstate for producing the strange particles (in a first approximation the two lowest octets) makes SU_3 appear as a secondary symmetry of dynamical origin.

In fact, if SU_3 were considered as a fundamental (i.e. exact) symmetry of the underlying natural law disturbed by an SU_3-asymmetrical vacuum, one would either expect Goldstone particles of mass zero and of undefined symmetry in SU_3, which interact strongly with the other particles, or one would have to assume a spectrum of the type studied by Higgs and Kibble, which both seem not to agree with the actual spectrum. Therefore it looks more natural to compare the SU_3-multiplets with the optical multiplets in the atomic spectra. These optical multiplets can be considered as the result of the group $O_3 \times O_3$ (independent rotation of the orbits and the spins of the electrons), which obviously is not a fundamental group; yet it may result approximately from dynamics in some parts of the spectrum.

On the other hand the classification of SU_3 as a secondary dynamical symmetry has consequences of considerable importance, which can be checked by the experiments. The most important one is the non-existence of quark-particles or of non integer electromagnetic charges. Only such (approximate) representations of SU_3, which can be obtained by multiplying the octet representation with another octet representation etc. could be expected as parts of the spectrum. Furthermore the algebra of the currents should represent very accurately $SU_2 \times SU_2$, but only with considerably minor accuracy $SU_3 \times SU_3$. All these results seem to be compatible with existing experimental evidence; but one may doubt whether or not future experiments could reveal the existence of quark-particles, which then would rule out Eqn. (1).

In any case the classification of SU_3 as secondary dynamical symmetry may be a controversial problem, and if there should be good arguments in favour of the opposite classification of SU_3 as fundamental symmetry, I hope that these arguments will be brought forward in the discussion.

Two more remarks should be added concerning the leptons and electrodynamics. In an approximation where the baryon octet is represented by one pole in (3), only the average lepton mass can be determined. If however the mass splitting in the octet is taken into account, there must be at least two different charged leptons corresponding in their symmetry to proton and Ξ. The particle connected with the proton gets a very small mass and should be identified with the electron, the other one gets a mass not very much smaller than the pion mass and corresponds to the muon. The muon mass has this rather high value in spite of being of purely electromagnetic origin. From this point of view one may call the leptons not 'real matter', but—as one might have done some hundred years ago—rather 'pure electricity' or 'electromagnetic singularities'.

Equation (1) has been interpreted mathematically along the lines of quantum electro-dynamics. Therefore it is not surprising that the mathematical scheme of quantum electro-dynamics seems to be contained completely in the more general scheme defined by (1). This can be seen in two ways. The operator $\chi^*(x)\sigma_\nu\chi(x)$ in (1) contains matrix elements connected with the creation or annihilation of photons (or more generally of electromagnetic field). If this part of the operator is called A_ν and treated separately from the rest of the operator, one recognizes in (1) the fundamental field equation of quantum electro-dynamics [the Dirac equation or more correctly Weyl equation for a (bare) electron without mass interacting with the electromagnetic field A_ν].

The other possibility is the translation of every special Feynman graph in quantum electro-dynamics into a corresponding graph in the formalism defined by (1). Any photon line in the electromagnetic formalism could in fact be replaced by the expression ⊂(η)⊃ of the non linear spinor theory. Hence e.g. the graph for electron–electron or proton–proton scattering

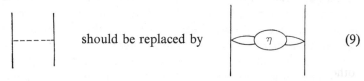
(9)

The latter graph contains, besides the electromagnetic forces, also short range forces which may be present besides the electromagnetic interactions, e.g. the proton–proton interaction by means of pions. Still the pole of ⊂(η)⊃ at mass zero guarantees the equivalence of the two graphs with respect to electromagnetic forces in that approximation in which the coupling constants on both sides are equal. Hence e.g. the well known formula for the Lamb shift expressed in terms of the charge and mass of the electron should be a consequence of (1) as well as of quantum electro-dynamics.

Almost nothing has been done in the non linear spinor theory so far with respect to the weak interactions. The violation of parity and the weakness of the interaction in β-decay obviously suggest that this phenomenon should, like the electromagnetic forces, be considered under the viewpoint of an asymmetry of the groundstate. The same should be true for the still smaller PC-violating interaction in K_2-decay and finally for gravitation. But no serious attempt has yet been made in this direction. As one argument in favour of an incorporation of weak interactions into a theory given by the Eqns. (1) and (3), one should mention the fact that the interaction term in (1) has just the same form as the interaction term in β-decay (especially with respect to the $(1 + \gamma_5)$-factor corresponding to a 2-component theory), and the universality of weak interaction.

3. Comparison with methods and results of other theories

Equation (1) contains a contact interaction which in the 'Feynman graphs' of the practical applications (5)–(7) appears as a vertex point, where four fermion lines meet. This fundamental interaction will in higher approximations lead to more indirect interactions which are produced by the exchange of other particles. In this way the formal scheme will in higher approximations gradually approach that kind of scheme which one would use in order to describe the assumptions of the bootstrap mechanism or other more phenomenological methods: Any particle may be considered as a compound system of other particles, held together by forces, which again are particles taken in the 'crossed channel'.

This is clearly seen in a recent paper by Dürr and F. Wagner, which investigates the baryon eigenvalue equation in an approximation where all wavefunctions with more than three variables are omitted. This approximation enables us already to consider the baryon as composed of one fermion and one boson, and should therefore include higher resonance states of any angular momentum. Actually it leads to an eigenvalue equation of the Bethe–Salpeter type, where the two particles, boson and fermion, are held together by an exchange force produced by the exchange of a fermion. Calculations of this type have been carried out on a phenomenological basis many years ago in a low energy approximation by Chew and Low, for the Bethe–Salpeter equations for scalar particles in case of mass zero exchange particles by Wick, Cutkowsky and others. The recent investigations of Dürr and Wagner show how the Tamm–Dancoff approximations gradually become similar to the more conventional schemes. Actually the Bethe–Salpeter equation seems to lead to a series of resonance states, where in each series the angular momenta increase by two units from one member to the next, in other words to Regge trajectories. The only difference against the more conventional schemes can be seen in the existence of the contact interaction and in the fact that some baryons can be expressed by the fundamental operator $\chi(x)$ alone, others only by the combination of at least three such operators. Whether this fact will finally have some influence on the shape of the Regge-trajectories is still an open question. The existence of the contact interaction may be connected in the bootstrap method with the problem of the behaviour of the dispersion integrals at very high energies.

The general dynamical aspects and the problems of convergence in the non linear spinor theory have recently been investigated in papers by Stumpf, Wagner, Wahl, Rampacher and others; in these papers the methods described in the first part of my talk have been applied to old problems of quantum mechanics, especially the anharmonic oscillator, in order to compare the various methods.

The numerical evaluation of a single step in the approximation procedure of the field theory has been simplified in a paper by Dürr and Wagner by the elaboration of the 'Feynman diagrams' to precise mathematical formulae. In a further development of this technique one may hope that an equation written thereby in the form of a 'Feynman diagram' can be directly translated into a programme for a computer. For the single integrals the computers have already been used by Géhéniau, Mitter and Biritz with considerable success. Hence a further extension of these methods seems quite feasible. It should be emphasized that a development in this direction will be necessary, independently of what finally the theoretical formulation of elementary particle physics will be. The degree of complication in the eigenvalue equations of this part of physics cannot possibly be smaller than that in the theory of complicated atoms or molecules, since in both cases we have typical many body problems. Therefore one will finally have to rely on computers in order to cope with this degree of complication. The adaptation of any method that has proved successful in many body physics will certainly also be useful.

The general tendencies in the bootstrap method as reported by Chew and in the non linear spinor theory are very closely related. However, at some points a theory starting from a master equation like Eqn. (1) will contain information which is not too easily added to the fundamental assumptions of the bootstrap formalism; yet they seem to be necessary to form a complete theory. The first point is that Eqn. (1) states the fundamental groups, which are—except for the Lorentz group—not stated explicitly in the bootstrap method. I cannot believe that the constraints in this formalism should suffice to determine the groups.

Then the analytical behaviour of the S-matrix elements is defined precisely by the differential Eqn. (1), while only general concepts like 'maximum analyticity' are available in the bootstrap scheme. Finally the concept of a degenerate vacuum, which is natural for a theory defined by (1), cannot without additive complications be incorporated into a pure S-matrix theory. Still, the general tendency of considering all particles as compound systems is common for both theories and is immediately suggested by the experiments.

Coming back to the general situation in the non linear spinor theory at the present time, it is clear that we are still very far from a complete theory on this basis. However, the outcome of the experiments during the last ten years may allow the conclusion that, when looking for a master equation for elementary particle physics, one need not look for anything more complicated than Eqn. (1).

May I make one final remark concerning a criticism which has often been expressed: that it is a too ambitious programme to try the formulation of a theory for the complete spectrum of elementary particles. Looking back on the development of atomic physics 40 years ago, it is clear that it would have been much more difficult to formulate a theory for only some part, say the triplet part, of the iron spectrum, than a theory for the complete spectrum. Therefore I feel that it would be more ambitious in our time to construct a theory only for the hadrons, or only for the leptons etc. than a theory for the complete spectrum.

Discussion on the report of W. Heisenberg

R. E. Marshak. As someone who is sympathetic to your programme to develop a unified theory of hadrons and leptons, I feel that your master equation in terms of one four-component field ('urmaterie') does not do full justice to the suggestiveness of the lepton triplet (v_e and $\bar{v}_\mu$ can be represented by one 4-component Dirac spinor and the other two are e and μ). Since the leptons are weakly interacting, they seem to be reflecting the attractiveness of starting with a master equation in terms of three fields (quark model). In your theory, you must work very hard to simulate broken SU_3 symmetry results (e.g. I am not sure that you can reproduce the Gell-Mann–Okubo formula for the baryon decuplet, the condition $2I + y = 0$ (mod. 2) etc.). Also, it would be difficult to understand the universality of the weak interactions between leptons and hadrons. Finally, if you do not believe one will observe 'urmaterie', I do not see why quarks should be found.

So much by way of comment. My question is whether in your theory, the muon mass has an electromagnetic origin. In the Baker–Johnson–Willey theory, the scale is not fixed for the electron mass but they hope to determine the (m_μ/m_e) ratio. So far they have not succeeded. I wonder how you can obtain such a large (m_μ/m_e) ratio by means of electromagnetism? My own inclination would be to connect m_μ with the hypercharge effect on mass for hadrons.

W. Heisenberg. First I would prefer to speak about 4 leptons (e, μ, v_e, v_μ) corresponding somehow to one-half of an octet; I cannot see any strong argument for speaking of a triplet and therefore in this sense referring to SU_3. But with respect to the question whether the muon mass is purely electromagnetic the situation in our theory is as follows: in that approximation, where the vacuum is isosymmetric (and when we therefore have no electromagnetic field), all lepton masses are zero. Hence they may afterwards all be called electromagnetic. When the vacuum is considered as asymmetric and when therefore the photons appear, those leptons that interact with the photon acquire their mass. But this mass is related to the baryon masses in the sense that each baryon needs supplementation (or you may call it compensation) by one (or possibly several) lepton dipoles, to make the photon of mass zero possible. The electron is the supplement of the photon, the muon that of the Ξ. In so far these masses are determined by the baryon masses, and the mass difference Ξ-proton has its counterpart in the different masses of e and μ.

With regard to the universality of weak interactions I would like to emphasize that it is at least a very natural basis for an understanding of this universality that baryons and leptons are created by the same field operator. The universality of electromagnetism (proton and positron have exactly the same charge) is due just to this feature of the theory. On account of the Ward identity the coupling constant (the charge) cannot depend on the amount of creation operator for a special kind of fermions, which is present in the universal field-operator. Therefore their charges are exactly equal. A similar situation could occur for the weak interactions, but this problem has not been worked out yet.

H. P. Dürr. I think we should not talk too much about leptons and weak interactions in connexion with the non-linear spinor theory at the moment. Too little has been done, up to now. Perhaps in ten years from now we will know a little more, and then we can come back to these points.

I want to comment on the possibility to consider $SU(3)$ as a fundamental symmetry broken only by the vacuum state. The first impression was that on the basis of the Goldstone theorem we would have to expect the occurrence of four scalar, strange massless particles if we only consider the violation of strong interactions. These scalar massless particles should interact strongly with hadrons and two of them should be charged. These particles we do not know experimentally, and we can be pretty certain that they do not exist. However, Higgs, Kibble, Brout, Englert and others have shown how the Goldstone theorem can be invalidated to some extent by coupling in gauge fields. In this case the strong breaking of $SU(3)$ would produce four massive strange vector mesons, so to say, half the octet. Generally speaking the spontaneous breakdown of a symmetry either produces mass zero particles or incomplete multiplets. Usually the four know strange vector mesons K^* are considered to be part of a vector–meson octet, including in addition the ρ and the ω. Hence one would have to find some other candidates for the incomplete multiplet to make such an interpretation acceptable.

I want to comment on another point. Heisenberg has compared the indefinite metric in the nonlinear spinor theory with the indefinite metric which occurs in the Gupta–Bleuler formalism of quantum electrodynamics. Now, it is rather clear that the indefinite metric in q.e.d. is of a much less dangerous type than the one used in the spinor theory, in particular, we can give a straightforward prescription how it can be avoided altogether. Nevertheless, there seems to be at least some formal similarity which I may express in the following way: in q.e.d. in the Bleuler–Gupta description we introduce two redundant fields, a longitudinal field connected with quanta of positive norm and a scalar field connected with quanta of negative norm. Taking the plus and minus combinations we obtain what I will shortly call a ghost couple, i.e. two ghost fields connected with quanta of zero norm, which, however, are not orthogonal to each other. One ghost, a 'good' ghost, obeys the Lorentz condition, the other, the 'bad' ghost, however, does not. We arrive at the physical theory by projecting out the 'bad ghost' by a subsidiary condition on the infields. Then the 'good ghosts' cannot do any harm because of their vanishing norm; they do not give contributions to any matrix elements and just reflect the gauge independence of the matrix elements. One can repeat the same construction for a massless spin two field couple to a conserved source function. Here one has to introduce 10 field operators where 8 fields are redundant. It can be shown that they constitute four ghost couples of the type described above. The four 'bad ghosts' are eliminated by the four Einstein conditions.

In a similar way the dipole ghost introduced in the nonlinear spinor theory constitute a ghost couple of this type, a 'good ghost' which is an energy eigenstate and a 'bad ghost' which is not. By the condition that physical states all must be eigenstates of the energy one can eliminate the bad ghosts, and hence arrive at a physically interpretable theory, provided that the bad ghosts do not produce bound states, which is hard to check.

E. C. G. Sudarshan. I find the equality of the electron charge and the proton charge to be extremely interesting and significant: it appears to be due to the use of the same description for both kinds of particles. Could you tell me how we can be assured that the electron or μ-meson would not get transmuted into a baryon? Needless to say the conservation of baryon number is also a very significant law which should not be violated without 'due processes of law'.

H. P. Dürr. The lepton conservation, I think, is a rather interesting point in our theory, but, I feel, also quite a hazardous one, which easily may prove to lack any basis. It actually states that the superselection rules we find in nature may not all arise from group theoretical conditions (gauge invariance) but also may arise from a peculiar analytical structure. In our case, baryons and leptons are not distinguished group theoretically, they transform under the same gauge transformation, and hence have the property to cancel each other's singularities. However, the leptons are dipoles and the baryons are poles in the Green's functions. Consequently the leptons, because of their vanishing norm (good ghost) have zero interaction with all particles, except with the photons, where the norm of the states is immaterial on the basis of Ward's indentity. Hence the baryon world is only linked to the lepton world by photons, and we therefore get a separate lepton and baryon conservation, if weak interactions later on do not mess up this separation. Of course, one has to be very careful about the statement that zero norm particles are coupled to photons. If we were to use Källén's approach to q.e.d. by employing a finite mass photon in which the zero limit is taken afterwards, then zero norm particles would not couple.

SIMPLEST DYNAMIC MODELS OF COMPOSITE PARTICLES

A. Tavkhelidze

Joint Institute for Nuclear Research, Dubna, U.S.S.R.

In our talk we wish to give a short review on recent work done by N. N. Bogoliubov and co-workers[1] concerning the study of electromagnetic and weak vertices of mesons and baryons in the simplest relativistic quark model.

As is well known these problems were investigated with success in the non-relativistic quark model. By the help of this model one could obtain some interesting results in connexion with the explanation of magnetic moments of mesons and baryons, radiative and weak decays, etc. It would be interesting to give an answer to the question concerned with the nature of the relativistic correction in the quark model.

Our paper consists of two parts, the first of which treats the semi-relativistic quark model, the second one being devoted to the method of construction of relativistic models.

1. Semi-relativistic model of quasi-independent quarks

In this model quarks in mesons or in baryons are considered as particles moving in a certain self-consistent field.

In this case the individual single-particle wave functions of each quark are meaningful. These functions are described by the Dirac equation for a particle in an external field

$$[\gamma_0 E - i\boldsymbol{\gamma}\cdot\boldsymbol{\nabla} - m - V(r)]\psi(r) = 0 \qquad (1.1)$$

where m indicates the free quark mass.

We assume that the potential $V(r)$ is spherically symmetric and that the lower energy level is in the S-state. Other properties of the potential are not essential for the model considered. Let us calculate the magnetic moment and the ratio of the axial and vector weak interaction constants of the nucleon.

Note first of all that baryons in this scheme are thought of as a three-quark system in the S-state. It is also required that the baryon wave function should be completely symmetric in spin and unitary spin indices. Owing to this requirement the baryon wave function transforms as the

56-plet of the $SU(6)$ group. Here, unlike the purely group approach, the orbital momentum of the quark is considered.

For the calculation of the magnetic moment of the nucleon the magnetic moment of the quark has to be determined. Postulating the minimal electromagnetic interaction for a free quark we get the following value for a renormalized magnetic moment of the quark

$$\mu_q = \frac{e_q}{2E_0}(1-\delta) \tag{1.2}$$

Here e_q denotes the charge of the quark; E_0 is the bound state energy, which by definition is equal to one third of the nucleon mass

$$E_0 = \tfrac{1}{3}m_N \tag{1.3}$$

δ is always positive-definite and is equal to the mean value of the L_z component of the orbital momentum in the states with total angular momentum $I_z = \tfrac{1}{2}$.

This formula is remarkable because the effective quark mass appears in the definition of the magnetic moment. Therefore an enhancement of the magnetic moment occurs in the scalar field.

Having obtained the formula for the magnetic moment of the quark the proton and neutron magnetic moments could be easily calculated. In terms of Bohr magnetons these quantities read

$$\begin{aligned}\mu_p &= 3(1-\delta) \\ \mu_n &= -2(1-\delta)\end{aligned} \tag{1.4}$$

Now we pass to calculate the weak interaction constants of the nucleon.

If the axial and vector constants of the free quarks are equal, for the renormalized values of these quantities we have

$$\begin{aligned}g_V &= g_0\tau^+ \\ g_A &= g_0(1-2\delta)\tau^+\sigma_z\end{aligned} \tag{1.5}$$

Here g_0 is the nonrenormalized weak interaction constant, while τ and σ are the isotopic and spin Pauli matrices. Hence, the ratio of the vector and axial constants for the nucleon reads

$$G_A/G_V = \tfrac{5}{3}(1-2\delta) \tag{1.6}$$

We see that the expression for the magnetic moments and weak interaction constants involves the quantity δ. Note that δ vanishes in the approximate nonrelativistic description when the lower components of the Dirac spinors are neglected. In this case for the proton and neutron magnetic moments we have

$$\mu_p = 3 \qquad \mu_n = -2 \tag{1.7}$$

while for the ratio of the axial and vector constants we obtain the well-known results of the $SU(6)$-symmetry:

$$G_A/G_V = \tfrac{5}{3} \tag{1.8}$$

For a tentative estimate we give here a value of δ which was calculated in a scalar potential well:

$$\begin{aligned} V(r) + m &= 0 \qquad r < r_0 \\ V(r) &= 0 \qquad r > r_0 \end{aligned} \tag{1.9}$$

when the large mass of the quark is completely compensated inside it. In this case $\delta = 0.17$ and

$$\mu_p = 2.49$$
$$G_A/G_V = 1.1$$

to be compared with experiment

$$\begin{aligned} (\mu_p)_{\exp} &= 2.79 \\ (G_A/G_V)_{\exp} &= 1.18 \end{aligned} \tag{1.10}$$

2. Attempts at relativistic generalization

Now we will be concerned with the problem of finding the vector and axial vertices of composite particle in the relativistic quark model.

Consider first a simplified model of mesons as a bound state of two spinless quarks. To describe this system it is possible to start from the relativistic invariant equation[1]

$$[\tfrac{1}{4}p^2 + q^2 - m^2]\varphi_p(q) - \int V_p(q, q')\,\delta(n \cdot q')\varphi_p(q')\,dq' = 0 \tag{2.1}$$

$$p \cdot q = 0 \qquad p^2 > 0$$

Here m is the quark mass, p is the four-momentum of the system, n_μ is the unit four-vector

$$n_\mu = \frac{1}{\sqrt{p^2}}\,p_\mu \tag{2.2}$$

The four-momentum q characterizes the intrinsic motion of the quarks.

The wave functions $\varphi_p(q)$ are normalized by the relativistic-invariant condition

$$\int \varphi_{p'}(q)\varphi_p(q)\,\delta(n\cdot q)\,dq = \delta_{MM'}$$

$$M = \sqrt{p^2} \qquad M' = \sqrt{p'^2} \tag{2.3}$$

In the absence of the interaction $V = 0$ it follows from (2.1) that for each quark the Klein–Gordon equation holds.

In the centre-of-mass system $\mathbf{p} = 0$ Eqn. (2.1) looks like

$$[\tfrac{1}{4}E^2 - \mathbf{q}^2 - m^2]\varphi_E(\mathbf{q}) - \int V_E(\mathbf{q},\mathbf{q}')\varphi_E(\mathbf{q}')\,d\mathbf{q}' = 0 \tag{2.4}$$

In this case the normalization condition reads

$$\int \varphi_{E'}(\mathbf{q})\varphi_E(\mathbf{q})\,d\mathbf{q} = \delta_{EE'} \tag{2.5}$$

As we can see the wave function φ_E in the centre-of-mass system does not depend on the relative energy q_0 or in the x-space it does not depend on the relative time t.

Later on, to determine in quantum field theory the vertex of the interacting quarks we use the following approach. Consider the one-time two-particle Bethe-Salpeter amplitude

$$\tilde{\varphi}_p(x,y) = \langle 0|\,\varphi(x)\varphi(y)\,|p\rangle_{x_0=y_0} = e^{ip(x+y/2)}\tilde{\varphi}_p(\mathbf{x}-\mathbf{y})\big|_{x_0=y_0} \tag{2.6}$$

$\varphi(x)$ denotes the second-quantized Heisenberg operator for the scalar field. It was shown[1] that the Fourier-transform of one-time wave function $\tilde{\varphi}_p(\mathbf{q})$ satisfies in the centre-of-mass system an equation of type (2.4) for the definite potential $\tilde{V}_E(\mathbf{q},\mathbf{q}')$.

It is essential to emphasize that there is a method of constructing this potential by means of Feynman graphs of the two-particle Green function.

Indeed, for the one-time wave function we have

$$\int d\mathbf{q}'\,\tilde{G}_E^{-1}(\mathbf{q},\mathbf{q}')\varphi_E(\mathbf{q}') = 0 \tag{2.7}$$

where the operator $\tilde{G}$ is the two-time four-particle Green function in the centre-of-mass system. By definition

$$\tilde{G}_E(\mathbf{q},\mathbf{q}') = \int dq_0\,dq_0'\,G_E(q,q') \tag{2.8}$$

Expanding the Green function as follows

$$G_i = G_0 + G_0 K G_0 + \cdots \tag{2.9}$$

where G_0 denotes a free Green function of two particles, and K is the Bethe-Salpeter kernel, for the inverse operator of the two-time Green function the relation is valid:

$$\tilde{G}^{-1} = \tilde{G}_0^{-1} + \widetilde{\tilde{G}_0^{-1} G_0 K G_0 \tilde{G}_0^{-1}} + \cdots. \tag{2.10}$$

The tilde denotes the integration as in (2.8).

Calculating $\tilde{G}_0$ in the centre-of-mass system we have

$$[\tilde{G}_0(\mathbf{q}, \mathbf{q}')]^{-1} = \frac{\pi}{i} \frac{\delta(\mathbf{q} - \mathbf{q}')}{\sqrt{\mathbf{q}^2 + m^2}} \cdot [\tfrac{1}{4}E^2 - \mathbf{q}^2 - m^2] \tag{2.11}$$

Hence, for the wave function $\tilde{\varphi}_E$, we get the equation

$$[\tfrac{1}{4}E^2 - \mathbf{q}^2 - m^2]\tilde{\varphi}_E(\mathbf{q}) - \frac{1}{\sqrt{\mathbf{q}^2 + m^2}} \cdot \int V_E(\mathbf{q}, \mathbf{q}')\tilde{\varphi}_E(\mathbf{q}')\, d\mathbf{q}' = 0 \tag{2.12}$$

The quasi-potential $\tilde{V}_E$ is given by the expansion

$$\tilde{V}_E = \widetilde{\tilde{G}_0^{-1} G_0 K G_0 \tilde{G}_0^{-1}} \tag{2.13}$$

In the framework of perturbation theory it is possible to check that the spectrum of bound states and the scattering matrix obtained with the help of Eqn. (2.12) coincides with similar quantities calculated on the basis of the Bethe–Salpeter equation.

Note that Eqn. (2.1) may be considered as a relativistic generalization of (2.12), provided that the following condition is fulfilled:

$$V_E(\mathbf{q}, \mathbf{q}') = \frac{1}{\sqrt{m^2 + \mathbf{q}^2}} \cdot \tilde{V}_E(\mathbf{q}, \mathbf{q}') \tag{2.14}$$

Let us now proceed to the case when the quarks have spin $\tfrac{1}{2}$. Here the quasipotential equation for the meson is written as

$$(\gamma_0^{(1,2)} E - \hat{M}_E)\psi = 0 \tag{2.15}$$

where $\hat{M}_E$ denotes the mass operator of the two interacting quarks.

$$\hat{M}_E \psi = 2\sqrt{m^2 + \mathbf{q}^2}\,\psi_E(\mathbf{q}) + \int \tilde{V}_E(\mathbf{q}, \mathbf{q}')\varphi_E(\mathbf{q}')\, d\mathbf{q}' \tag{2.16}$$

The normalization condition is

$$\int d\mathbf{q}'\, \psi_{E'}(\mathbf{q}')\psi_E(\mathbf{q}') = \delta_{EE'} \tag{2.17}$$

In finding the vertex functions of the bound states in terms of quasi-potential wave functions one faces difficulties. In the presence of the external fields the momentum of the initial and final states are not, generally speaking, conserved. So it is impossible to define the rest system. Therefore, we restrict ourselves to constructing the vector and axial charges, that is, the vertex functions for the zero momentum transfer. For this purpose it is convenient to use the following method.

Let us switch on the external field which does not carry energy and momentum, but can carry other quantum numbers. Such fields were treated by Wentzel and they are usually called spurion fields. The interaction of quarks with spurion fields are chosen as

$$\Gamma^{(i)} = \begin{cases} \gamma_\mu^{(i)} \lambda_\alpha^{(i)} a^{\mu\alpha} & \text{for vector spurions} \\ \gamma_5^{(i)} \gamma_\mu^{(i)} \lambda_\alpha^{(i)} a_5^{\mu\alpha} & \text{for axial spurions} \end{cases} \quad (2.18)$$

$a^{\mu\alpha}$, $a_5^{\mu\alpha}$ denote the constant C-numbers which define the amplitudes of the vector and axial spurion fields with quantum numbers of the unitary octet.

Then the quasipotential equation in the spurion field takes the following form

$$[\gamma_0^{(1,2)} E - \hat{M} + \delta\hat{M}]\psi = 0 \quad (2.19)$$

where $\hat{M}$ is the mass operator in the centre-of-mass system. The operator $\delta\hat{M}$ represents the renormalization of the bound state mass in the spurion field. It can be found by the perturbation theory expansion:

$$\delta\hat{M} = \overline{G}^{-1} \overline{\delta G_a} \, \overline{G}^{-1} \quad (2.20)$$

Here the operator δG_a denotes the first variation of the two-particle Green function $G_p(q, q'; a)$ in the external spurion field and is equal to a sum of the following diagrams

$$\begin{array}{c}(2) \\ (1)\end{array} \; \underline{\quad} + \underline{\cdot\quad\cdot} + \underline{\quad\xi} + \underline{\xi\quad} + \cdots$$

that is, diagrams for the two-particle Green function with one spurion vertex. The upper sign in this formula denotes

$$\overline{\delta G}_E(\mathbf{q}, \mathbf{q}'; a) = \int dq_0 \, dq_0' G_E^F(q, q') \quad (2.21)$$

where the integrand is taken in the Foldy–Wouthysen representation. On the basis of Eqn. (2.19) for the effective charges of the composite particle we obtain the following formula

$$\int \psi_E(\mathbf{q}) \, \delta\hat{M} \psi_E(\mathbf{q}) \, d\mathbf{q} = Q_{\mu_\alpha} a^{\mu\alpha} + Q_{5\mu_\alpha} a_5^{\mu\alpha} \quad (2.22)$$

Note that the explicit form of the charges depends essentially on the form of the mass operator $\hat{M}$.

When there is no interaction between quarks it is possible to get the following expression for the vector and axial charges

$$\langle 0| Q_\alpha |0\rangle = \int \psi_E(\mathbf{q})[\lambda_\alpha^{(1)} + \lambda_\alpha^{(2)}]\psi_E(\mathbf{q})\, d\mathbf{q}$$

$$\langle 0| Q_{5\alpha} |0\rangle = \int \psi_E(\mathbf{q})[\lambda_\alpha^{(1)}\Delta_z^{(1)} + \lambda_\alpha^{(2)}\Delta_z^{(2)}]\psi_E(\mathbf{q})\, d\mathbf{q}$$

(2.23)

where

$$\Delta_z^{(1)} = \sigma_z^{(1)} \frac{m}{\sqrt{m^2+\mathbf{q}^2}} + q_z \frac{\boldsymbol{\sigma}^{(1)} \cdot \mathbf{q}}{\sqrt{m^2+\mathbf{q}^2}(\sqrt{m^2+\mathbf{q}^2}+m)}$$

One can see from this formula that the vector charge of two non-interacting quarks is not renormalized and is equal to a sum of charge of individual quarks.

The axial charge of a two-quark system is, however, renormalized even in the absence of the interaction between them. For example, in the case of spherical symmetric wave functions ψ from (2.23) we have

$$\langle 0| Q_{5\alpha} |0\rangle = (\lambda_\alpha^{(1)}\sigma_z^{(1)} + \lambda_\alpha^{(2)}\sigma_z^{(2)})\left(1 - \frac{\langle \mathbf{q}^2\rangle}{3m^2} + \cdots\right) \quad (2.24)$$

Now we compare this expression with the renormalized values of the axial charges of the quarks which were calculated earlier in the quasi-independent quark model. Expanding δ in powers of $1/m$ gives

$$\delta = \frac{\langle \mathbf{q}^2\rangle}{6m^2} + \cdots \quad (2.25)$$

It can be easily seen that the first-order calculated values for the renormalized axial constants for both models coincide. The point is that in the expansion (2.25) the first term does not depend on the interaction potential and is only due to the relativistic effects.

The method described is closely connected with the programme suggested by Gell-Mann for constructing representations of the local algebra of currents by means of the relativistic quark model[2].

To describe the meson Gell-Mann[2] uses the Yukawa–Markov type equation of the form

$$(\hat{p} - \hat{M})\psi = 0 \quad (2.26)$$

with the renormalization condition for the wave function

$$\int \bar{\psi}_{p'}(q)\psi_p(q)\, dq = \delta_{MM'}$$

Eqn. (2.26) is an analogue of Eqn. (2.1) if quarks have spin $\frac{1}{2}$. From Lorentz-invariance the most general form of the matrix element of the current is written as

$$\mathscr{F}_{\alpha\mu}(p, p') \sim \int dq \, dq' \, \sqrt{\delta(p \cdot q) \, \delta(p' \cdot q')} \, \bar{\psi}_{v'}(q') G_{\alpha\mu} \psi_v(q) \quad (2.27)$$

where $\hat{G}_{\mu\alpha}$ is a sum over $\lambda_\alpha^{(1)}$ and $\lambda_\alpha^{(2)}$ times all possible independent vector operators such as $\gamma_\mu^{(1)}$, $\gamma_\mu^{(2)}$ etc, times Lorentz-invariant functions

$$F(q \cdot p', p \cdot q' \cdots q^2, q'^2)$$

A requirement is imposed on the invariant functions so that the currents thus determined should satisfy the algebra in the system $p_z \to \infty$. Such a possibility in the model of non-interacting quarks was illustrated in Gell-Mann's lectures.

In conclusion we point out that there exists a relationship between the relativistic quark model and the group theoretical approach using the infinite component wave equations.

If we assume that the interaction between quarks takes place by exchange of scalar massless particles then in the instantaneous interaction Eqn. (2.12) becomes a Schrödinger equation in the Coulomb field.

As is known all the solutions of this equation form the basis of one unitary infinite-dimensional representation of the group $O(4.1)$.

Therefore the problem of determining the energy levels of the bound states can be reduced to finding unitary representation of some non-compact group. At present there exists no method for constructing the potential. Therefore, by analogy with the Coulomb problem, attempts are being made to describe the bound states of the system by means of infinite-component wave equations, that is, the problem of determining the potential is replaced by that of finding the symmetry group. The substitution of dynamics by group theory has been one of the recent trends in elementary particle physics. Having seen a general qualitative success of the quark model of mesons and baryons, one wants a more qualitative, but simple and unified description of hadron phenomena without assuming detailed dynamical mechanisms.

References

(1) A. Logunov and A. Tavkhelidze, *Nuovo Cimento*, **29**, (1963).
(2) M. Gell-Mann, Lectures at the International School of Physics 'Ettore Majorana', Erice, Sicily (1966).

Discussion on the report of A. N. Tavkhelidze

H. P. Dürr. There is a point which I have never understood in the dynamical treatment of the quarks: whether they are fermions or bosons. If one succeeds in some way or another in describing the nucleon as a three quark bound state by introducing, for example, a scalar interaction, as you have done, isn't the crucial question here whether the four, five and particularly the six quark state (the deuteron?) etc. have a binding energy compatible with experiment?

A. N. Tavkhelidze. It is obvious that the 3 quark state is of special importance since 2, 4 or 5 quark states have not been observed. The deuteron should be considered most likely to be a system consisting of two 3 quark subsystems. From the dynamical point of view, it is difficult to explain the exceptional role of the 3 quark system.

L. A. Radicati. (1) Did you obtain a particular representation of Gell-Mann's local algebra or a class of representations?

(2) The second question is not specifically directed to you but to anybody in the audience who knows how to answer. Is it obvious that a representation of Gell-Mann's algebra should coincide with a representation of a non-compact group? Is this a general property?

A. N. Tavkhelidze. (1) We have obtained only the simplest representation of the local algebra. The problem of the existence of the representations of other types has not been considered.

(2) The fact that the representation of the local algebra coincides with the representation of the non-compact group is, in may opinion, very interesting and surprising. At least, I do not know any mathematical theorems which would explain this coincidence.

F. E. Low. In the Hartree field model $\Sigma \varepsilon_i \neq E$; one has to add in the potential. Therefore, in your one particle theory, one must explicitly add the potential, which means that E_0 again involves the quark mass. Therefore it is hard to take the specific numerical results seriously.

A. N. Tavkhelidze. Starting from the qualitative arguments about the nature of the forces which bind quarks, and using the independent quark model, it is possible to calculate such quantities as magnetic moments, axial constants and the effective masses of quarks. Collective effects can be taken into account by means of perturbation theory. It is obvious that the equations which describe the residual interactions of bounded independent quarks must include the quark effective mass m_{eff} and not the large real mass M. In any case the calculations performed by means of the

Bethe–Salpeter equation would be extremely useful if we know better the nature of the forces acting between quarks.

W. Heisenberg. What are the symmetry properties of the quarks in your model? Since you assume the spin $\frac{1}{2}$ for the quarks, one should consider Fermi statistics. On the other hand you spoke about the wave function being symmetrical in the quarks. What are your assumptions?

A. N. Tavkhelidze. The totally symmetric spin-unitary spin wave function gives a good description of the main experimental properties of baryons. However, if quarks are fermions such a choice contradicts the Pauli principle. To avoid this difficulty attempts were made to introduce a totally antisymmetric wave function of three quarks in the s-state or to ascribe to quarks the parastatistics. There is also a possibility of introducing instead of one triplet of fractionally charged quarks three triplets of integer charged quarks, the main consequences of the $SU(6)$ symmetry being unaffected.

DISSIPATIVE PROCESSES, QUANTUM STATES AND FIELD THEORY

I. Prigogine*†
Université Libre de Bruxelles, Belgium.

Synopsis

This report is devoted to the study of quantum states and quantum levels corresponding to a finite lifetime. Closely connected is the question of a consistent dynamical description involving physical (renormalized) particles. Reasons are discussed (see §1) why it is convenient to use the density matrix as a starting point. The general methods of non-equilibrium statistical mechanics are summarized in §2. For further reference the interaction between radiation and matter is discussed in §3 and the difficulties to include higher order terms in the coupling constant summarized in §4.

A central role in this entire report is played by a new transformation theory (see §5). It leads (through a non unitary transformation) to a new description in which the dynamical evolution is especially simple and physically clear. It is very similar to the evolution of a weakly coupled system. It is important that at each moment the entropy of the system can be expressed through a simple Boltzmann type combinatorial expression. This guarantees that we are dealing with physical (or dressed particles) in the sense of field theory.

States both with infinite lifetime and finite lifetime can be defined. In the first case they are defined by a wave function, but in the second case they are not (since they are not in the Hilbert space). The theory is applied to the interaction of radiation and matter (§6). It is shown that Einstein's classical theory of spontaneous transitions can be generalized in this way to higher order in the coupling constant. The explicit expression of an excited quantum state and quantum level including lifetime effects are discussed in connexion with a simple example in §7.

Relation with field theory is discussed in §8. For stable particles the usual mass and charge renormalization is recovered. However for

* This report was prepared with the collaboration of Dr. Cl. George and Dr. F. Henin.
† Also the Center for Statistical Mechanics and Thermodynamics, University of Texas, Austin, Texas.

unstable particles new effects appear. The main difference is that no renormalized field operator can be defined. To illustrate our method the neutral scalar field with fixed sources and the Lee model is discussed (see §§9–10). It is shown that our method provides us with a generalization of improper unitary transformations for dissipative systems. Reasons are given why ghost states do not appear in our theory.

The relation between quantum states and entropy takes a remarkable new form (§11).

Dressing processes in dissipative systems appear as essentially irreversible processes with increase of entropy. This feature is retained in our theory which leads to a dynamical description which unifies in a sense general features of statistical mechanics and of field theory.

1. Introduction*

The similarity of certain types of problems which exist in field theory and in statistical mechanics has been often stressed. In both cases, we deal with systems with an infinite number N of degrees of freedom (or at least we have to study the limit $N \to \infty$). Therefore some type of asymptotic procedure is unavoidable to extract physical information from such systems. More precisely the parallel of a free field is a model of harmonic oscillators such as may be used as a model for classical solids at low temperatures while coupled fields correspond to interacting systems of statistical mechanics as they must include some mechanism to allow for sufficiently long times (in a finite volume) the establishment of thermodynamical equilibrium.

Now great progress has been achieved in the time description of the systems usually considered in statistical mechanics, such as interacting molecules in a gas or interacting normal modes [Prigogine (1962, 1967), Balescu (1963), Résibois (1966)].

Therefore it seems natural to consider some field theoretical problems from this perspective. While I am well aware of the lack of generality of the results presented in this paper,† I feel however that they may point to an interesting new direction. Indeed they seem to indicate the possibility of a new formulation of field theory which is much more similar to statistical mechanics. For this reason I have prepared this report, in which the emphasis will be on physical ideas while the mathematical developments (which are elementary but often lengthy) will be found elsewhere.

---

* The author apologizes for the unusual length of this paper. He wanted to provide not only a readable account of the statistical formalism used but also to include a few examples to give the reader a feeling of the physical meaning of the concepts involved.

† See footnote on p. 202.

[Henin, Prigogine, George, Mayné (1966), Prigogine and Henin (1967), George (1967), Prigogine, Henin and George (1968).]

Many important problems of elementary particle physics may be discussed in terms of the S-matrix formalism once the existence of asymptotic in and out states is *assumed*. However, in non-equilibrium statistical mechanics one deals *always* with an initial state which is by definition a non-equilibrium state and cannot be extended to $t \to -\infty$. Similarly unstable particles cannot be included in the asymptotic in-states.

Clearly if we could define precisely what we mean by a quantum state with *finite* lifetime we could generalize the S-matrix approach and discuss the transition probability P_{if} where i or f, or both, are defined without introducing the asymptotic condition $t \to \pm\infty$. This will be the main problem we shall discuss.

Let us first discuss a few simple examples in which excited states contribute to the thermodynamic properties and the scattering cross section.

We consider a system characterized by the Hamiltonian

$$H = H_0 + \lambda V \qquad (1.1)$$

In subsequent paragraphs, we shall often consider the example of an atom with a single electron interacting with a radiation field. It is therefore appropriate to give the explicit form of H_0 and V for this case. In the nonrelativistic limit the Hamiltonian is then the sum of the Hamiltonian H_0^a of the atom and of H_0^f which is the free transverse field Hamiltonian; V describes the interaction between the atom and the transverse photon field. The eigenstates and eigenvalues of H_0^a will be labelled $|\mu\rangle$ and ε_μ. We shall also introduce occupation numbers n_μ for these levels. We may then write

$$H_0^a = \sum_\mu n_\mu \varepsilon_\mu \quad (\text{with } \sum_\mu n_\mu = 1 \quad \text{and} \quad n_\mu = 0 \text{ or } 1) \qquad (1.2)$$

Similarly we may introduce photon occupation numbers $\{n_\kappa\}$, and we have

$$H_0^f = \sum_k n_k \omega_k \qquad (1.3)$$

We restrict ourselves to the term linear in the charge $[-e(\mathbf{P} \cdot \mathbf{A}/m)]$ in the interaction potential V, so that we may write

$$\lambda V = \sum_{k\mu\nu} \sum_{\varepsilon=\pm 1} V^{(\varepsilon)}_{\mu/\nu k} a_\mu^\varepsilon a_\nu^{-\varepsilon} a_k^{-\varepsilon} \qquad (1.4)$$

where a^{+1}, a^{-1} are the usual creation and destruction operators. A term such as

$$a_\mu^{+1} a_\nu^{-1} a_k^{-1} \qquad (1.5)$$

corresponds to a process in which a photon k is destroyed and the atom makes a transition from level ν to level μ. We neglect two photon processes which can be treated by the same method.

The explicit form of $V^{(e)}_{\mu/\nu k}$ will not be necessary here.

We shall also later consider, as examples, the scalar model of field theory (see §9) as well as the Lee model (see §10).

Let us first consider the thermodynamic equilibrium properties of an atom together with the radiation field when enclosed in some box of finite volume. We *assume the validity of Gibbs canonical formalism at equilibrium*. Therefore we have for the partition function

$$Z = \text{tr } e^{-H/kT} = \sum_n (e^{-H/kT})_{nn} \tag{1.6}$$

where the n are occupation numbers [see (1.2), (1.3)]. If the interaction is sufficiently weak (1.6) reduces to the much simpler expression

$$Z = \sum_n e^{-(H_0)_{nn}/kT} \tag{1.7}$$

In this approximation the thermodynamic properties are expressed in terms of the occupation numbers of the photons and the electron. However in this approximation the levels are calculated neglecting the finite life time of the excited states (as they are eigenfunctions of the time displacement operator H_0).

On the contrary (1.6) involves the levels of the system as a whole. We could also calculate in this case the average number of photons present but this calculation would not be meaningful as we do not know how to distinguish between real and virtual photons.

The question is now if we can write (1.6) *even in the presence of the interaction* in the form

$$Z = \sum_n e^{-E(\{n\})/kT} \tag{1.8}$$

where the energy levels $E(\{n\})$ may be still functionals of the occupation numbers. We would then have something like effective energy levels which take into account the fact that excited levels have only a finite lifetime.

There is one important physical example where this has been done: namely in Landau's theory of Fermi liquids (see Nozières, 1963). However, in this case, we are interested in states near the Fermi surface and deal with a non dissipative or weakly dissipative situation.

The question is therefore: can we extend Landau's theory to dissipative systems? Interesting results have been obtained by Balian and de Dominicis (1960). However their method seems to be strictly restricted to equilibrium and will not be employed here*.

* A detailed comparison between the kinetic method we use and the method of Balian and de Dominicis is in preparation (see footnote on p. 202).

Another very interesting attempt is due to Résibois (1965a). [See also Watabe and Dagonnier (1966); for earlier attempts to introduce renormalization in the frame of kinetic theory see Prigogine and Leaf (1959), Henin (1963).]

To lowest order in the coupling constant the method we use in this paper reduces in the case of Fermi particles to Résibois' results.

Let us now consider a simple scattering problem. We consider an incident photon ω_λ which is scattered by a 'two-level' atom and observe the outgoing photon ω_f. We may of course apply usual S-matrix scattering theory (see the excellent textbook by Goldberger and Watson, 1964).

Fig. 1.1. Resonance scattering.

The characteristic feature of this problem is that we may consider at least two possible mechanisms for the scattering: either we form first, through absorption of the initial photon ω_λ, the excited state b which is then deactivated through emission of ω_f, or level b enters only 'virtually' in the process. In the first case we expect to find in the corresponding cross section the product of δ-functions,

$$\delta(\varepsilon_b - \varepsilon_a - \omega_\lambda)\, \delta(\varepsilon_b - \varepsilon_a - \omega_f) \qquad (1.9)$$

while in the second case we expect the single δ-function,

$$\delta(\omega_\lambda - \omega_f) \qquad (1.10)$$

Moreover, in the first case the duration of the process involves the lifetime of the excited state b and may be expected to be much longer (for matter–radiation interaction) than in the second case.

In the situation for which ω_λ is far from $\varepsilon_b - \varepsilon_a$, the process (1.9) cannot occur. We may then calculate the scattering matrix by the lowest order Born approximation which involves the two Feynman diagrams (Figure 1.2.). In this way one obtains a well defined contribution [see Grosjean (1959), Goldberger and Watson (1964), Kroll (1965)] which we don't need to reproduce here.

We inquire what happens when

$$\omega_\lambda \to \varepsilon_b - \varepsilon_a \qquad (1.11)$$

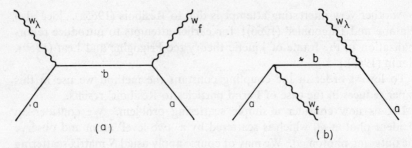

Fig. 1.2. Feynman diagrams corresponding to the Born approximation for resonance scattering.

Then the T-matrix becomes singular since, in the diagram (a) of Figure 1.2., we have the propagator

$$\frac{1}{\varepsilon_a + \omega_\lambda - \varepsilon_b + z} \tag{1.12}$$

which in the limit $z \to 0$ has a pole on the real axis. To avoid this one replaces the propagator by the complete Green's function corresponding to the intermediate state b. In terms of the so-called level shift operator R (see Goldberger and Watson, 1964) this gives instead of (1.12) the expression

$$\frac{1}{\varepsilon_a + \omega_\lambda - \varepsilon_b + z - \langle b| R(z) |b\rangle} \tag{1.13}$$

and the poles are shifted into the lower half plane. Neglecting diagram (b) of Figure 1.2., and taking the square of the contribution (a) with the lifetime correction expressed by (1.13), one obtains, introducing some minor approximations, the well known Lorentz shape. However it should be noticed that in the limit (1.11) and $z \to 0$ we obtain a contribution of order

$$1/\lambda^2 \tag{1.14}$$

where λ is the coupling constant.

We have therefore avoided the divergence problem but we have to give up any expansion in powers of the coupling constant. As the result the precision of this method is difficult to discuss. It does not even lead to an expression valid for the whole range of frequencies: the Lorentz shape is only derived when diagram (b) is neglected, which is only possible near resonance, while the Born approximation is obtained when resonance is neglected. More complicated situations may be discussed in a similar way (for the standard treatment see Kroll, 1965).

Can we do better? In all transport equations discussed for quantum mechanical systems such as the Pauli equation or the Boltzmann equation for dilute gases we deal directly with *transition probabilities* and not with transition amplitudes. However in this approximation the transition probabilities can be easily calculated in terms of the T-matrix since in this case any transition $i \to f$ can only be achieved in a single way. As an example we shall discuss in detail the Boltzmann (or Pauli) equation for the matter-radiation interaction in §3.

Our problem is therefore to derive 'higher order' transport equations with the hope of isolating the different processes such as (1.9) (1.10) and to discuss their interplay in given physical situations.

What we need is an extension of the *Einstein treatment of spontaneous emission* (1917) based on transition probabilities to higher order in the coupling constant. But this involves a clear and precise definition of the excited level ε_b in (1.9).

Let us study these problems starting from the von Neumann equation for the density matrix

$$i \frac{\partial \rho}{\partial t} = [H, \rho] \qquad (1.15)$$

In this way we are able to make direct contact with non-equilibrium statistical mechanics. But there is a more fundamental reason why (1.15) is an appropriate starting point. While Schrödinger's equation is linear in the wave function, the wave function appears quadratically in average values. Now in all problems we shall discuss here there are asymptotic elements involved (limit of large volume, long time $\cdots$). In the T-matrix method these asymptotic elements are introduced first and then the square is taken while in methods involving the statistical density matrix such as the kinetic methods one works directly with quadratic functionals of the wave functions.

As the density matrix satisfies both a linear equation and is also linearly related to average values it is not astonishing that asymptotic procedures are in fact simpler to handle. Let us now briefly summarize the method which leads to the kinetic equations we shall use.

2. Kinetic equations

There are at present very beautiful and compact methods to derive from (1.15) the so-called master equation (see specially Résibois 1966, Zwanzig 1960, 1964, Balescu 1967, Baus to appear). We shall try to give a concise account of the physical ideas involved.

In terms of ρ the average value of an observable A may be written as

$$\langle A \rangle = \sum_{nn'} \langle n| \rho |n'\rangle \langle n'| A |n\rangle \qquad (2.1)$$

It is useful to perform a change of variables

$$n - n' = \nu; \quad n + n' = 2N \tag{2.2}$$

and to use the notation

$$\langle n| A |n'\rangle = A_{n-n'}\left(\frac{n+n'}{2}\right) = A_\nu(N) \tag{2.3}$$

In this way (2.1) becomes

$$\langle A \rangle = \sum_{\{N\}} \{A_0 \rho_0 + \sum_\nu A_\nu \rho_\nu\} \tag{2.4}$$

In a model in which a random phase approximation would be valid all ρ_ν would vanish. It is therefore appropriate to consider ρ_ν as expressing the correlations in the system while ρ_0 refers to the '*vacuum of correlations*'*.

The aim of the theory is to establish exact equations of evolution for the vacuum component ρ_0 and the 'correlation component' formed by all the ρ_ν (with $\nu \neq 0$). To do this, we first write (1.15) using the notation of (2.2) and (2.3) together with (1.1). After a few elementary manipulations, we then obtain

$$-i\frac{\partial \rho_\nu(N,t)}{\partial t} + \langle \nu| L_0 |\nu\rangle \rho_\nu(N,t) = \lambda \sum_{\nu'} \langle \nu| \delta L |\nu'\rangle \rho_{\nu'}(N,t) \tag{2.5}$$

with

$$\langle \nu| L_0 |\nu'\rangle = e^{(-\nu'/2)(\partial/\partial N)}(H_0)_{\nu-\nu'}e^{(\nu/2)(\partial/\partial N)} - e^{(\nu'/2)(\partial/\partial N)}(H_0)_{\nu-\nu'}e^{-(\nu/2)(\partial/\partial N)}$$

$$= (\sum_\mu \nu_\mu \varepsilon_\mu + \sum_k \nu_k \omega_k)\delta_{\nu\nu'} \tag{2.6}$$

Similarly

$$\langle \nu| \delta L |\nu'\rangle = e^{(-\nu'/2)(\partial/\partial N)}(V)_{\nu-\nu'}e^{(\nu/2)(\partial/\partial N)} - e^{(\nu'/2)(\partial/\partial N)}(V)_{\nu-\nu'}e^{(-\nu/2)(\partial/\partial N)} \tag{2.7}$$

The two terms in (2.6) and (2.7) correspond to the two terms in the original commutator (1.15).

The interest of (2.5) lies in its striking analogy with the Liouville equation of classical statistical mechanics. The variables ν play the role of the classical angle variables while the N play the role of the action variables. For this reason we have called (2.5) the 'Liouville–von Neumann' equation. Since (2.5) describes 'transitions' between the vacuum component ρ_0 and the correlation component ρ_ν, or transitions *inside* the correlation components we may say that (2.5) describes the *dynamics of correlations*. We

---

* This distinction can be made in a specially clear and elegant way by introducing projection operators (see the references at the beginning of this paragraph).

shall not write the explicit form of (2.7) but notice that because of the form of the potential energy (1.4) the only non vanishing elements are

$$\langle v| \, \delta L \, |v_\mu + \varepsilon, v_{\mu'} - \varepsilon, v_\kappa - \varepsilon \rangle \quad \text{with} \quad \varepsilon \pm 1 \tag{2.8}$$

This may be represented by a vertex at which two electron lines and one photon line meet. Examples are given in Figure 2.1. While the graphical expression is very similar to the Feynman graph, Figure 1.2, the physical meaning is quite different. For example, as shown by formula (2.15) the matrix element

$$\langle 0| \, \delta L \, |1_\mu, -1_{\mu'}, -1_k \rangle \, \rho_{1_\mu, -1_{\mu'}, -1_k}$$

gives a contribution to ρ_0. In other words the matrix elements of δL correspond to transitions between different correlations (or between

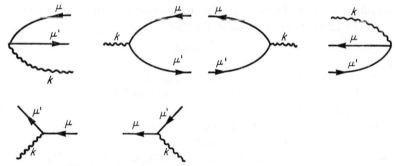

Fig. 2.1. Graphical expression of matrix elements (2.8).

correlations and the vacuum of correlations) while Feynman diagrams correspond to transitions between states. Also as indicated by (2.7) the matrix elements $\langle v| \, \delta L \, |v' \rangle$ are still operators acting on the occupation numbers N.

These differences are not unexpected: Feynman's graphs are associated with the evolution of the wave function while our diagrammatic representation refers directly to the evolution of the density matrix (which is quadratic in the wave function).

The Liouville–von Neumann equation can be solved formally using various tools such as Laplace transforms or projection operators to obtain the evolution equations for the vacuum component ρ_0 as well as for the correlation component ρ_ν.

For ρ_0 we obtain the evolution equation valid both in the classical and the quantum cases

$$\frac{\partial \rho_0}{\partial t} = \int_0^t d\tau \, G(t - \tau) \rho_0(\tau) + \mathscr{D}(t; \rho_\nu(0)) \tag{2.10}$$

In this equation the quantity $G(t)$ is a generalized collision operator defined formally in terms of all irreducible vacuum of correlations to vacuum of correlations transitions (see 2.11). A fundamental role in the theory is played by the Laplace transform $\psi(z)$ of $G(t)$ with

$$\psi(z) = \langle 0 | \, \delta L \left(\sum_{n=1}^{\infty} \frac{1}{z - L_0} \delta L \right)^n | 0 \rangle_{\text{irr}} \qquad (2.11)$$

where the index 'irr' means that only terms such that all intermediate states are different from the vacuum of correlations $|0\rangle$ must be taken into account.

Again the formal similarity with the irreducibility condition used in the T-matrix or level shift operator should be emphasized (see Goldberger and Watson, 1964). However, there this condition refers to intermediate quantum states while here it refers to intermediate states of correlations.

In the subsequent development we shall require

$$\lim_{z \to +i0} \psi(z) \qquad (2.12)$$

The irreducibility condition in (2.11) has the consequence that $\psi(z)$ has no poles on the real axis. Using the theory of Cauchy integrals the limit (2.12) may be taken in an unambiguous way [see Henin, Prigogine, George and Mayné (1966), Prigogine (1967), Schieve (1967)].

That is the basic property which makes an expansion of the collision operator in powers of the coupling constant possible. The finite duration of the collision is expressed through the non-instantaneous character of Eqn. (2.10). The second term in the r.h.s. of (2.10) expresses the influence of initial correlations.

The fundamental asymptotic property which makes this formulation important can be summarized by the following assumption

$$\mathscr{D}(t; \rho_v(0)) \to 0 \qquad (2.13)$$

For simple cases this property may be verified by direct calculation. It should be stressed that it can only be satisfied for very large systems and for well defined classes of initial distributions corresponding to correlations of a finite range.

The entire foundation of thermodynamics is included in (2.13). If this condition were not satisfied the system could not forget its initial conditions and would not evolve to the state of maximum entropy independently of the initial conditions.

The property (2.13) will also be used in the field theoretical problems we shall discuss. In fact, it is the assumption which will permit us to introduce well defined *physical* particles. As we shall see in more detail later it provides a deep link between ergodicity and the problem of elementary particles.

If we assume (2.13) and moreover use (2.11), we may rewrite (2.10) in the form

$$i\frac{\partial \rho_0}{\partial t} = \psi\left(i\frac{\partial}{\partial t}\right)\rho_0(t) \tag{2.14}$$

The lowest approximation to (2.14) is obtained by neglecting the time dependence in the operator ψ as

$$i\frac{\partial \rho_0(t)}{\partial t} = \psi(0)\rho_0(t) \tag{2.15}$$

This is possible when one has two widely separated time scales t_{coll} and t_{rel} such that

$$\frac{t_{\text{coll}}}{t_{\text{rel}}} \to 0 \tag{2.16}$$

where t_{coll} is a time characterizing the duration of the collision process and t_{rel} is the relaxation time of the process. The duration of the collision is considered as instantaneous on the scale of the relaxation time*. We may say that (2.15) corresponds to the Boltzmann approximation of statistical physics.

A more general equation is obtained when $\psi[i(\partial/\partial t)]$ is formally developed in a power series of $[i(\partial/\partial t)]$. After reordering the series, Eqn. (2.14) becomes

$$i\frac{\partial \rho_0(t)}{\partial t} = \Omega\psi(0)\rho_0(t) \tag{2.17}$$

where Ω is a functional of $\psi(z)$ and its derivatives in respect to z, for $z \to +i0$. For example the first terms are

$$\Omega = 1 + \psi'(0) + \tfrac{1}{2}\psi''(0)\psi(0) + [\psi'(0)]^2 + \cdots \tag{2.18}$$

with

$$\psi'(0) = \left(\frac{d\psi}{dz}\right)_{z \to +i0}, \qquad \text{etc} \cdots. \tag{2.19}$$

The basic difference between (2.15) and (2.17) is that in (2.17) the finite duration of the collision is taken into account through the operator Ω.

The transition from Eqn. (2.14) to Eqn. (2.17) is only possible if the relaxation process still corresponds to the *longest* relevant time scale in the system. Therefore there are important problems involving many degrees of freedom which cannot be described through (2.17); examples are gravitational interactions [see Prigogine and Severne (1966), (1967)] and spin relaxation through a Heisenberg Hamiltonian [see Résibois and

* For more details concerning the time scales involved, see the references given in the introduction.

De Leener (1966)]. Such 'non-Boltzmannian' situations are excluded explicitly from our subsequent discussion. We shall call (2.17) the 'post-Boltzmannian' approximation [see Prigogine and Henin (1967), Prigogine (1967)].

To conclude this survey let us consider briefly the question of correlations. Any given correlation may be split into two parts:

$$\rho_v(t) = \rho_v'(t) + \rho_v''(t) \tag{2.20}$$

The evolution of the first part is given by an equation similar to the kinetic equation for $\rho_0(t)$ which describes the scattering of the free correlations. This part vanishes for $t \to \infty$ exactly as (2.13). The second part corresponds to the creation of correlation from $\rho_0(t)$. It is given by

$$\rho_v''(t) = \int_0^t d\tau \mathscr{C}_v(\tau) \rho_0(t - \tau) \tag{2.21}$$

where the 'creation fragment' is the Laplace transform of the operator

$$\mathscr{C}_v(z) = \langle v| \sum_{n=1}^{\infty} \left(\frac{1}{z - L_0} \delta L \right)^n |0\rangle_{\text{irr}} \tag{2.22}$$

The index 'irr' again means that no intermediate state is identical to the vacuum of correlations. Expanding $\rho_0(t - \tau)$ around $\rho_0(t)$ we obtain $\rho_v''(t)$ in terms of the distribution function $\rho_0(t)$ at the same time. We then find an expression of the form

$$\rho_v''(t) = C_v \rho_0(t) \tag{2.23}$$

This corresponds exactly to the post-Boltzmannian approximation (2.17) in which we also have expressed $\partial \rho_0 / \partial t$ in terms of ρ_0 taken at the same time.

Both Eqn. (2.10) for the vacuum component and (2.28) for the correlation component may be obtained from a single equation by taking the 'projections' of this equation on the vacuum and on the correlations [see especially Zwanzig (1960), Balescu (1967)]. Here we have considered homogeneous situations in the absence of external forces but the method we have used may be generalized to apply to other situations [Severne (1965), Balescu (1967)].

As an example let us consider the Boltzmannian and the post-Boltzmannian approximation for the interaction between radiation and matter.

3. Boltzmann approximation for the interaction between radiation and matter

Let us start with (2.15) and use the general expression (2.11) for the collision operator. We are interested in the lowest order which is λ^2

(corresponding to the Born approximation). In our diagram representation (see Figure 2.1) this corresponds to two vertex diagrams connecting the vacuum of correlations to the vacuum of correlations. The form of these diagrams is indicated in Figure 3.1. These diagrams correspond to the formation of a correlation involving a single photon with a lifetime which is of the order of the duration of the collision. Using the explicit form of

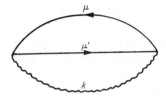

Fig. 3.1. Contribution to the collision operator to order λ^2.

the Hamiltonian (see 1.1–1.4) as well as of the matrix elements of the Liouville operator one obtains the operator

$$-i\psi_2(0) = -2\pi \sum_\mu \sum_{\mu'} \sum_k \sum_{\varepsilon=\pm 1} \delta(\varepsilon_\mu - \varepsilon_{\mu'} - \omega_k) |V_{\mu/\mu'k}|^2$$

$$\times \delta_{N_\mu,(1+\varepsilon)/2} \delta_{N_{\mu'},(1-\varepsilon)/2} \left(N_k + \frac{1+\varepsilon}{2} \right)$$

$$\times \left[1 - \exp\left\{ -\varepsilon \left(\frac{\partial}{\partial N_\mu} - \frac{\partial}{\partial N_{\mu'}} - \frac{\partial}{\partial N_k} \right) \right\} \right] \quad (3.1)$$

This operator describes single photon absorption or emission processes with a corresponding change of state of the electron. It is still an operator acting on the occupation numbers N. Therefore the kinetic Eqn. (2.15) takes the explicit form

$$\frac{\partial \rho_0(N, t)}{\partial t} = -2\pi \sum_{\mu\mu'} \sum_k \delta(\varepsilon_\mu - \varepsilon_{\mu'} - \omega_k) |V_{\mu/\mu'k}|^2$$

$$\times \left\{ \delta_{N_\mu,1} \delta_{N_{\mu'},0} (N_k + 1) \left[1 - \exp\left(-\frac{\partial}{\partial N_\mu} + \frac{\partial}{\partial N_{\mu'}} + \frac{\partial}{\partial N_k} \right) \right] \right.$$

$$\left. + \delta_{N_\mu,0} \delta_{N_{\mu'},1} N_k \left[1 - \exp\left(\frac{\partial}{\partial N_\mu} - \frac{\partial}{\partial N_{\mu'}} - \frac{\partial}{\partial N_k} \right) \right] \right\} \rho_0(N, t)$$

$$= -2\pi \sum_{\mu\mu'} \sum_k \delta(\varepsilon_\mu - \varepsilon_{\mu'} - \omega_k) |V_{\mu/\mu'k}|^2 \{ \delta_{N_\mu,1} \delta_{N_{\mu'},0}$$

$$\times (N_k + 1)[\rho_0(N) - \rho_0(\{N\}', N_\mu - 1, N_{\mu'} + 1, N_k + 1)]$$

$$+ \delta_{N_\mu,0} \delta_{N_{\mu'},1} N_k [\rho_0(N) - \rho_0(\{N\}', N_\mu + 1, N_{\mu'} - 1, N_k - 1)] \}$$

$$(3.2)$$

This equation is, of course, well known and often called the 'Pauli-equation'. We recognize in the r.h.s. the familiar gain and loss terms. The transition probabilities which appear in the Pauli equation correspond to the Born approximation for one photon process. For future reference let us summarize the main properties of this weakly coupled approximation:

(1) The operator $i\psi_2(0)$ given by (3.1) is hermitian. It may indeed be verified that

$$\sum_N f_i^*(N) i\psi_2(0) f_j(N) = [\sum_N f_j(N) i\psi_2(0) f_i^*(N)]^* \qquad (3.3)$$

This is in fact a general property valid to all orders of λ (see George, to appear). Therefore the eigenvalues of this operator are real. The evolution of the system may be described in terms of real relaxation times.

(2) There is an obvious correspondence in (3.1) between the δ-function

$$\delta(\varepsilon_\mu - \varepsilon_{\mu'} - \omega_k) \qquad (3.4)$$

which expresses conservation of the (unperturbed) energy and the displacement operator

$$\exp\left[-\varepsilon\left(\frac{\partial}{\partial N_\mu} - \frac{\partial}{\partial N_{\mu'}} - \frac{\partial}{\partial N_k}\right)\right] \qquad (3.5)$$

both referring to the same three-particle process.

(3) For long times we have [for the proof see Prigogine (1962), Résibois (1966)]

$$\rho_0 \to \sim \exp\left[-\frac{H_0}{kT}\right] \qquad (3.6)$$

We obtain the unperturbed canonical distribution. This is a special case of a general rule: if in the kinetic equation terms up to order λ^{2n} are retained, the macroscopic properties are correct up to order λ^{2n-2}. Since we retain only terms of order λ^2 we cannot obtain any effect related to correlations (which are at least of order λ^2). In our specific case the asymptotic equilibrium distribution corresponds, for a finite temperature, to a Boltzmann distribution of matter and a Planck distribution of radiation.

(4) As all correlations are neglected the general formula for averages (2.4) reduces to the sum taken over the diagonal

$$\langle A \rangle = \sum_N A_0(N) \rho_0(N) \qquad (3.7)$$

(5) The entropy (or the $\mathcal{H}$-quantity) is given both for equilibrium and out of equilibrium by the Boltzmann functional

$$\mathcal{H} = \sum_N \rho_0(N) \log \rho_0(N) \qquad (3.8)$$

It is also a sum taken over the diagonal.

(6) The relation

$$\varepsilon_i = \frac{\partial H_0}{\partial N_i} \tag{3.9}$$

between the Hamiltonian of the system (as a whole) and the one-particle energies which enter in the kinetic equation, is trivially satisfied. All these properties are well known. We have however repeated them in order to discuss the deep alterations introduced when we go to higher order approximations.

We may directly apply (3.2) to various problems of physical interest. For example, to discuss spontaneous emission we must start with the atom in the excited state and no photons present in the system. This is expressed by the initial condition

$$\rho_0(1_2, t = 0) = 1 \quad \text{all other } \rho_0 = 0 \tag{3.10}$$

We assume therefore that the atom was initially in the second excited level. Equations (3.2) give us the usual exponential decay.

Let us use the abbreviations

$$\gamma_{\mu/\nu k} = 2\pi \, \delta(\varepsilon_\mu - \varepsilon_\nu - \omega_k) |V_{\mu|rk}|^2 \tag{3.11}$$

$$\gamma_\mu = \sum_{\nu k} \gamma_{\mu/\nu k} \tag{3.12}$$

The only diagonal elements which remain for times much longer than the life times γ_2, γ_1 of the excited states are

$$\rho_0(1_0, 1_k, t)\underset{t\to\infty}{=} \frac{\gamma_{2/0k}}{\gamma_2} \tag{3.13}$$

$$\rho_0(1_0, 1_k, 1_{k'}, t)\underset{t\to\infty}{=} \frac{\gamma_{2/1k}\gamma_{1/0k'} + \gamma_{2/1k'}\gamma_{1/0k}}{\gamma_1\gamma_2} \tag{3.14}$$

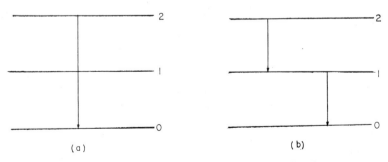

Fig. 3.2. Spontaneous emission in the Born approximation.

Clearly (3.13) corresponds to emission of a single photon due to the quantum transition 2–0 (Figure 3.2) while (3.14) corresponds to the production of two photons corresponding to successive transitions from 2 to 1 and from 1 to 0 (Figure 3.2b). The average photon number is given by

$$\langle n_k \rangle_{t \to \infty} = \sum_{\{N\}} N_k \rho_0 = \frac{\gamma_{2/0k}}{\gamma_2} + \frac{\gamma_{2/1k}}{\gamma_2} + \frac{\gamma_{1/0k}}{\gamma_1} \sum_{k'} \frac{\gamma_{2/1k'}}{\gamma_2} \quad (3.15)$$

It is formed by three sharp lines corresponding to the δ-functions involved in $\gamma_{2/0k}$, $\gamma_{2/1k}$, and $\gamma_{1/0k}$. We may study *resonance scattering* in the same way. We take as our initial condition

$$\rho_0(1_0, 1_\lambda, t = 0) = 1: \quad \text{all other } \rho_0 = 0 \quad (3.16)$$

The atom is initially in the ground state and a photon λ is present. We now have the absorption process followed by emission processes with the atom returning to the ground state. If as usual in scattering theory we retain only terms in $1/L^3$ we may neglect sequences corresponding to more than a single absorption process each of which gives a factor L^{-3}. One then obtains after a few elementary calculations again for times long with respect to the lifetime of the excited states*

$$\rho_0(1_0, 1_\lambda, t)_{t \to \infty} = 1 - \lambda^2 t [\gamma_{1/0\lambda} + \gamma_{2/0\lambda}] \quad (3.17)$$

$$\rho_0(1_0, 1_k, t)_{t \to \infty} = \lambda^2 t \left\{ \frac{\gamma_{1/0\lambda}\gamma_{1/0k}}{\gamma_1} + \frac{\gamma_{2/0\lambda}\gamma_{2/0k}}{\gamma_2} \right\} \quad (3.18)$$

$$\rho_0(1_0, 1_k, 1_{k'}, t)_{t \to \infty} = \lambda^2 t \frac{\gamma_{2/0\lambda}[\gamma_{2/1k}\gamma_{1/0k'} + \gamma_{2/1k'}\gamma_{1/0k}]}{\gamma_1 \gamma_2} \quad (3.19)$$

The results have the same interpretation as in the case of spontaneous emission: we have either elastic scattering described by (3.18) or inelastic scattering (3.19) with the emission of two photons. In this approximation we have scattering only when the resonance conditions corresponding to the δ-functions are satisfied. Therefore there is no finite line width. This is of course very unrealistic and shows that we have to go beyond the Boltzmann approximation, but here the difficulties begin!

4. The post-Boltzmannian approximation

We start now with Eqn. (2.17) taken together with (2.18). If we include terms up to order λ^4 we obtain

$$\frac{\partial \rho_0}{\partial t} = -i\lambda^2 \psi_2 \rho_0 - i\lambda^4 \psi_4 \rho_0 - i\lambda^4 \psi'_2 \psi_2 \rho_0 \quad (4.1)$$

---
* To simplify the notation we restrict ourselves to a 3-level system.

We now have to take into account contributions to the collision operator involving four vertices. Examples of such diagrams are indicated in Figure 4.1. The characteristic feature of the contributions represented in Figure 4.1 is that they describe correlations involving two photons.

Fig. 4.1. Contributions to the collision operator to order λ^4.

It is a straightforward but lengthy calculation to obtain the explicit form of ψ_4 [see for a similar calculation Henin, Prigogine, George, Mayné (1966)]. We shall not reproduce it here but we wish to make the following qualitative remarks (following essentially the same order as in §3) in order to compare the Boltzmannian and post-Boltzmannian approximations.

(1) The operator $i\psi_4$ is still hermitian, also ψ_2' is hermitian. However, the r.h.s. of (4.1) contains the product of two non-commuting hermitian operators which act on the occupation numbers

$$\psi_2'(i\psi_2)$$

Therefore the collision operator in (4.1) is no longer hermitian and we cannot conclude *a priori* that the approach to equilibrium will be monotone;

(2) When the calculations are performed we find different types of δ-functions and displacement operators. In addition to (3.4) and (3.5) we find also

$$\delta(\omega_k - \omega_{k'}) \tag{4.2}$$

$$\delta(\varepsilon_\mu - \varepsilon_{\mu'} - \omega_k \pm \omega_{k'}) \tag{4.3}$$

and

$$\exp\left[-\varepsilon\left(\frac{\partial}{\partial N_k} - \frac{\partial}{\partial N_{k'}}\right)\right] \tag{4.4}$$

$$\exp\left[-\varepsilon\left(\frac{\partial}{\partial N_\mu} - \frac{\partial}{\partial N_{\mu'}} - \frac{\partial}{\partial N_k} \pm \frac{\partial}{\partial N_{k'}}\right)\right] \tag{4.5}$$

Unfortunately, a given δ-function is no longer associated with the 'correct' displacement operator (for example 4.2 may be associated with 4.5). This is not a phenomenon peculiar to the matter-radiation interaction studied. It has been noticed by Mangeney (1964) for a plasma interacting with a radiation field [see also Baus (1966), Chapell, Britten and Glass (1965)] and by Résibois (1963, 1965, 1966a) for collisions in a moderately dense system. This prevents us from any simple interpretation of Eqn. (4.1) in terms of energy conserving collision processes;

(3) Still it may be shown [see Prigogine (1962), Prigogine and Henin (1960)] that for long times the correct canonical equilibrium distribution is recovered

$$\left.\begin{array}{c}\rho_0\\ \rho_v\end{array}\right\}_{t\to\infty} \to \text{canonical distribution} \qquad (4.6)$$

As a special case Eqn. (4.1) leads to a correct equilibrium up to order λ^2.

(4) In the evolution of macroscopic quantities correlation effects have to be taken into account; formula (3.7) is no longer valid.

(5) Correlation effects must also be taken into account in the entropy; the Boltzmann expression (3.8) is no longer valid [for a more detailed discussion of this point, see Prigogine, Henin and George (1966)].

(6) Of course relation (3.9) is no longer valid.

We see therefore that the theory of strongly coupled systems as described by the post-Boltzmannian approximation differs radically from the theory of weakly coupled systems. While the theory is mathematically consistent, the physical interpretation of the effects involved is not clear*.

Perhaps this increased complexity is not unexpected. Here the interactions play a double role: on the one hand they permit the system, as it does in weakly coupled systems, to reach equilibrium, but on the other hand they modify the asymptotic state which now includes the effect of interactions [see (4.6)]. This double role of the interactions leads to a greatly increased complexity. It is therefore quite remarkable that it appears possible, at least in the frame of perturbation technique, to write the kinetic equations for a strongly coupled system in a representation in which the simplicity of the weakly coupled case is preserved. We wish now to indicate briefly how this can be done.

5. Transformation theory

We have already stressed in §1 the importance of dressing and renormalization and mentioned some of the difficulties one has to face if one tries to introduce these ideas into a theory of dissipative systems such as those which involve excited atoms or unstable elementary particles.

First of all it should be stressed that a unitary transformation [as used e.g. in quantum field theory, see Heitler (1954)] would be of no interest for

---

* We could say that no particle interpretation is possible as real and virtual processes are mixed.

us. Indeed the form of the kinetic Eqn. (2.17) which corresponds essentially to an iterated commutator [see for more details Résibois (1963a)] would remain unaltered and all the differences with weakly coupled systems we have enumerated in §5 would persist*. Therefore a mere change in the constants involved in the Hamiltonian (so called mass and charge renormalizations) would be of no assistance here.

We start again with Eqn. (2.17) whose formal solution may be written

$$\rho_0(t) = e^{-i\Omega_\psi t}\bar{\rho}_0(0) \tag{5.1}$$

where $\bar{\rho}_0(0)$ is a 'redefined' initial condition. Since we neglect short time transients in the asymptotic Eqn. (2.17) we have to redefine the initial condition to secure the correct *long time* behaviour. A simple calculation given elsewhere [Prigogine and Henin (1967), Prigogine (1967), George (1967)] shows that the relation between $\bar{\rho}_0(0)$ and the exact initial condition $\rho_0(0)$† may be expressed in terms of $\psi(z)$ and its derivatives for $z = 0$. One finds, indeed, that

$$\bar{\rho}_0(0) = A\rho_0(0) \tag{5.2}$$

with

$$A = \sum_{n=0}^{\infty} \frac{1}{n!} \left[\frac{d^n\{\psi(z)\}^n}{dz^n} \right]_{z \to +i0} \tag{5.3}$$

An important feature of A is that it is a hermitian operator. Indeed the quantity $(i\psi^{(n)}(0)$ is hermitian for n even and antihermitian for n odd). The operator A has many remarkable properties which are discussed elsewhere (George, 1967, to appear).

It appears as a kind of projection operator projecting the real initial conditions $\rho_0(0)$ into a subspace of the space in which the distributions $\rho_0(t)$ are defined.

Combining (5.1) and (5.2) we obtain

$$\rho_0(t) = e^{-i\Omega_\psi t}A\rho_0(0) \tag{5.4}$$

This expression contains two non-commuting operators Ω_ψ and A and may be written in an alternative form which is of great interest here. Indeed we have also the 'symmetrical' form

$$\rho_0(t) = \chi e^{-i\varphi t}\chi\rho_0(0) \tag{5.5}$$

* A unitary transformation could only be useful if the terms in λ^4 in (4.1) would all vanish. This seems completely out of the question.
† In order to simplify the notations we assume here that initially all correlations $\rho_\nu(0)$ vanish.

where χ is a time independent operator. By identification with (5.4) we obtain

$$\chi\chi = A \tag{5.6}$$

$$\varphi = \chi^{-1}\Omega\psi\chi \tag{5.7}$$

The great interest of this formal manipulation is that $i\varphi$ is hermitian exactly as is the collision operator $i\psi$, while $\Omega\psi$ is not! This may be easily verified to lowest order in the coupling constant. Using (5.3) and (5.6) we obtain

$$\chi = 1 + \frac{\lambda^2}{2}\psi'(0) \tag{5.8}$$

and therefore (5.7) becomes

$$i\varphi = i\lambda^2\psi_2(0) + i\lambda^4\{\psi_4(0) + \tfrac{1}{2}[\psi_2'(0)\psi_2(0) + \psi_2(0)\psi_2'(0)]\} \tag{5.9}$$

Now the hermitian operators $\psi_2'(0)$, $i\psi_2(0)$ appear together in a *symmetrized way and therefore* $i\varphi$ is hermitian. Let us introduce the new distribution function

$$\tilde{\rho} = \chi^{-1}\rho_0 \tag{5.10}$$

We see immediately from (5.5) that $\tilde{\rho}$ satisfies the kinetic equation

$$i\frac{\partial \tilde{\rho}}{\partial t} = \varphi\tilde{\rho} \tag{5.11}$$

with the hermitian collision operator $i\varphi$. We have therefore already achieved an interesting result: the evolution of $\tilde{\rho}$ as given by (5.11) will be of the same nature as for weakly coupled systems. But this property does not define χ completely. It can be shown (George, 1967), that (5.5) can be generalized in the form

$$\rho_0(t) = \chi e^{-i\varphi t}\chi^+\rho_0(0) \tag{5.5'}$$

with

$$\chi = \chi'\chi'' \tag{5.12}$$

where χ' is a hermitian operator given by (5.6) and χ'' an unitary operator. In this way $i\varphi$ still remains hermitian.

How are we to choose this unitary operator? For this we go back to the expression of an average value (2.4). If we use also (2.20) (2.21) as well as (2.23) and neglect memory effects included in ρ_v' we obtain

$$\langle A \rangle = \sum_N \left(A_0\rho_0 + \sum_{v\neq 0} A_{-v}C_v\rho_0\right)$$

$$= \sum_N \left(A_0 + \sum_{v\neq 0} C_v^+ A_v\right)\rho_0 \tag{5.13}$$

where C_ν^+ is the adjoint operator corresponding to C_ν. We write this briefly as

$$\langle A \rangle = \sum_N (A_0 + C^+ A') \rho_0 \qquad (5.14)$$

This formula is of the same type as (4.7) valid for a weakly coupled system. This simple result can only be valid for large dissipative systems. If this were not the case, we could not neglect initial correlation effects and express offdiagonal elements of ρ in terms of the diagonal ones. We would then also be unable to express the average value of A in terms of the diagonal elements ρ_0 *alone*.

It is very interesting to notice that the large number of degrees of freedom and the dissipativity of the system have been used to simplify considerably the description of the system*. Using (5.10) we obtain

$$\langle A \rangle = \sum_N \chi^+ (A_0 + C^+ A') \tilde{\rho}$$
$$= \sum_N \mathscr{A} \tilde{\rho} \qquad (5.15)$$

It can be seen that if we associate with every quantum mechanical operator A the operator $\mathscr{A}$ such that†

$$\mathscr{A} \equiv \chi^+ (A_0 + C^+ A') \qquad (5.16)$$

we may calculate averages exactly as for the case of weakly coupled systems, with ρ_0 being replaced by $\tilde{\rho}$. As a special case we have for the energy

$$\langle H \rangle = \sum_N \chi^+ (H_0 + C^+ V) \tilde{\rho}$$
$$= \sum_N H_R(N) \tilde{\rho}(N) \qquad (5.17)$$

with the definition of the 'Hamiltonian' associated with $\tilde{\rho}$ given by

$$H_R = \chi^+ (H_0 + C^+ V) \qquad (5.18)$$

There still remains a difference in that we have yet no relation between $\tilde{\rho}$ and the entropy. If we could find a $\tilde{\rho}$ such that, exactly as in (3.8),

$$\mathscr{H} = \sum_N \tilde{\rho}(N) \log \tilde{\rho}(N) \qquad (5.19)$$

---

* This is of course well known in kinetic theory of gases where the existence of dissipation (collisions) is the foundation on which the existence of normal solutions of the Boltzmann equation is based.

† We consider however here only the class of observables defined in a homogeneous system.

then we would have a theory of strongly coupled systems which in all essential aspects, would be formally identical to the theory of weakly coupled systems. The effects of the interaction would then be included in the *definitions* of the energy and the entropy (5.15), (5.19) as well as in the form of the kinetic operator. The main idea is now to use the undetermined unitary operator χ'' (see 5.12) to obtain precisely the relation (5.19) between entropy and distribution function.

If such a relation exists, the maximization of $(-\mathcal{H})$ together with (5.17) would lead at equilibrium to the distribution function

$$\tilde{\rho}_{t\to\infty} \sim \exp\left(-\frac{H_R}{kT}\right) \tag{5.20}$$

Now to each χ operator there corresponds a different form (5.3) of the collision operator and the equilibrium distribution has to be an eigenfunction of the collision operator corresponding to an eigenvalue zero

$$\varphi \exp\left(-\frac{H_R}{kT}\right) = 0 \tag{5.21}$$

Both φ and H_R are functionals of χ'' and therefore this condition may be used to *construct the explicit form of* χ''. This has been done up to order λ^6 in the kinetic operator for a class of problems including the problem of interaction between matter and radiation. A single unitary χ'' operator satisfying this condition has been found (Prigogine and George, to appear; Prigogine, Henin and George, to appear). This representation of the kinetic equation which makes it look similar to the Boltzmann approximation may be called the *Boltzmann representation*.

At this point, let us make two remarks: first, as has been shown by George (1967) the effective Hamiltonian H_R may be written as

$$H_R = \chi^{-1} H_0 \tag{5.22}$$

This is exactly the same relation as the relation (5.10) between $\tilde{\rho}$ and ρ_0. Inversely, it can be shown (George, 1967) that (5.10) may be written in a form similar to (5.16)

$$\tilde{\rho} = \chi^+(\rho_0 + D\rho') \tag{5.23}$$

In this way $\tilde{\rho}$ appears as a combination between the diagonal elements ρ_0 and the correlations ρ'. We may say (George, 1967) that $\tilde{\rho}$ leads to a 'contracted' statistical description of the system. Instead of a description both in terms of the vacuum component ρ_0 and of the correlation component ρ_ν, we have a *single* quantity $\tilde{\rho}$ which combines the effect of both and describes the evolution of the system in terms of a single equation.

As the correlational effects are included in the definition of $\tilde{\rho}$ it is not astonishing that we have now a situation which is so closely parallel to that for weakly coupled systems. But the entities which are associated with the new Hamiltonian H_R and described by the kinetic Eqn. (5.11) are of course no longer the *initial* entities but *new entities*. In §8 we shall indicate why we can really consider these new entities as the '*physical*' particles (or dressed particles) in the usual sense of field theory.

Let us first discuss the meaning of quantum mechanical states and of energy levels in this representation. The quantum mechanical state of a system in which one electron at time $t = 0$ is on the excited level $|\mu\rangle$ would be now described in terms of $\tilde{\rho}$ by

$$\tilde{\rho}(1_\mu) = 1 \tag{5.24}$$

This specification corresponds, in terms of the initial distribution function, to the diagonal elements

$$\rho_0 = \chi\tilde{\rho}(1_\mu) \tag{5.25}$$

and to off diagonal elements (see 2.23)

$$\rho_\nu = C_\nu[\chi\tilde{\rho}(1_\mu)] \tag{5.26}$$

In the initial representation this state corresponds to a statistical mixture. This is general as long as the Boltzmann representation can be defined: an excited state of an atom, a molecule, an unstable elementary particle corresponds to a '*pure state*' in the sense of (5.24) in the Boltzmann representation but to a mixture in the representation corresponding to bare states.

We shall consider later a few examples (excited atom in §7, scalar model in §9, Lee model in §10). The important point is that this mixture cannot be reduced by any canonical transformation to a pure state in the Hilbert space. In other words an excited state (or any state with finite lifetime) cannot be represented in terms of wave functions in the Hilbert space. On the contrary if we have a situation in which the atom is in the ground state

$$\tilde{\rho}(1_0) = 1 \tag{5.27}$$

this condition can be shown to be equivalent to the condition that the atom is in the state described by the *wave function* corresponding to the (physical) ground state. We have, therefore,

$$\tilde{\rho} = |\tilde{\Psi}^+\rangle\langle\tilde{\Psi}| = 1 \tag{5.28}$$

We shall see an example in §10.

We have, therefore, two types of quantum states, according as they can or cannot be defined in terms of the usual Hilbert space. Moreover we may now calculate the energy levels corresponding to the different excited levels by applying (3.9) in the form

$$\tilde{\varepsilon}_i = \frac{\partial H_R}{\partial N_i} \tag{5.29}$$

Of course, exactly as in the Landau theory of Fermi liquids, these levels may still depend on the occupation numbers.

Note that the levels are real and sharp. The time evolution as due to the kinetic operator is completely separated from the definition of the energy levels. Again for the ground state, or more generally in the absence of dissipative effects, the usual formulae are recovered. But in the presence of dissipative effects new expressions are derived (see for examples §§7–10). It is worthwhile to insist how far we are from the usual quantum mechanical description: no longer is a state the eigenfunction of some operator and the energy levels its eigenvalues! We obtain new quantum rules. All this is due to the dissipative effects: when they may be neglected as it is the case near the Fermi surface the usual Landau theory is recovered.

Before we go into a deeper discussion of these new rules, let us first consider briefly the problem of interaction between radiation and matter.

6. The post-Boltzmannian approximation for the interaction between radiation and matter

We shall now apply the transformation theory summarized in §6 to the interaction between radiation and matter. As an illustration we shall mainly discuss the Wigner–Weisskopf model (Heitler, 1954). This corresponds to a two-level atom with the supplementary condition that

$$V_{1/0k} \neq 0 \quad \text{but} \quad V_{1k/0} = 0 \tag{6.1}$$

As the result of this simplification the physical ground state coincides with the bare ground state at zero temperature.

The kinetic equation (5.11) may be written including terms to order λ^6

$$i\frac{\partial \tilde{\rho}}{\partial t} = (\lambda^2 \varphi_2 + \lambda^4 \varphi_4 + \lambda^6 \varphi_6)\tilde{\rho} \tag{6.2}$$

The operator φ_2 is exactly of the same structure as the Boltzmann operator (3.1). It contains a δ-function

$$\delta(\tilde{\varepsilon}_1 - \tilde{\varepsilon}_0 - \omega_k) \tag{6.3}$$

associated with the displacement operator

$$\exp\left[-\varepsilon\left(\frac{\partial}{\partial N_1} - \frac{\partial}{\partial N_0} - \frac{\partial}{\partial N_k}\right)\right] \quad (\varepsilon = \pm 1) \tag{6.4}$$

The only difference is that the energy levels and the square of the interaction energy $|V_{\mu/\nu k}|^2$ are modified. We now have the 'renormalized' transition probability (see 3.11)

$$\tilde{\gamma}_{1/0k} = 2\pi\delta(\tilde{\varepsilon}_1 - \tilde{\varepsilon}_0 - \omega_k)|\widetilde{V_{1/0k}}|^2 \tag{6.5}$$

and

$$\tilde{\gamma}_1 = \sum_k \tilde{\gamma}_{1/0k} \tag{6.6}$$

We shall discuss briefly the meaning of the renormalization later in this paper. The operator φ_4 corresponds to elastic scattering of the photon and is given by

$$-i\varphi_4 = 2\pi \sum_{k_1 k_2} \delta(\omega_{k_1} - \omega_{k_2})\sigma_4(k_1 k_2)(N_{k_1} + 1)N_{k_2}\left[1 - \exp\left(\frac{\partial}{\partial N_{k_1}} - \frac{\partial}{\partial N_{k_2}}\right)\right]$$

$$\tag{6.7}$$

In this process one photon is destroyed and another emitted. The δ-function does not contain the atomic states. We may therefore state (see §1 of this paper) that the excited atomic states enter only virtually in this transition probability. The explicit expression of $\sigma_4(k_1, k_2)$ is, (apart from renormalization effects)

$$\sigma_4(k_1, k_2) = P\frac{1}{\tilde{\varepsilon}_1 - \tilde{\varepsilon}_0 - \omega_{k_1}}\left(\frac{\partial}{\partial \omega_{k_1}} + \frac{\partial}{\partial \omega_{k_2}}\right)|V_{1/0k_1}|^2|V_{1/0k_2}|^2 \tag{6.8}$$

As usual the notation

$$P\frac{1}{x} \tag{6.9}$$

means the principal part of $1/x$.

In the general case of an arbitrary number of levels we would find in φ_4 other processes such as Raman effect and simultaneous emission (or absorption) of two photons. But we shall neither write these terms nor indicate here the explicit form of φ_6. The main point is that now all terms which are appearing in the collision equation have the correct association between δ-functions and displacement operators and may be understood as transition probabilities. Let us now treat briefly spontaneous emission and resonance scattering in the framework of the Wigner–Weisskopf model.

To study spontaneous emission we start with [see (3.10)]

$$\rho_0(1_1) = 1, \quad \text{all other } \rho_0 = 0 \text{ for } t = 0 \tag{6.10}$$

The atom is initially in the *unperturbed excited* state. We may then calculate the initial state $\tilde{\rho}$ in the quasiparticle distribution using (5.10) and then proceed with the kinetic equation (5.11) exactly as was done in §3. We obtain correct to order λ^2 (for simplicity we neglect here the effect of φ_6)

$$\tilde{\rho}(1_0, 1_k, t)_{t\to\infty} = \left\{ \frac{\tilde{\gamma}_{1/0k}}{\tilde{\gamma}_1} - \lambda^2 P \frac{1}{\tilde{\varepsilon}_1 - \tilde{\varepsilon}_0 - \omega_k} \frac{\partial}{\partial \omega_k} |V_{1/0k}|^2 \right\}$$

$$\times \left\{ 1 + \lambda^2 \sum_{k'} P \frac{1}{\tilde{\varepsilon}_1 - \tilde{\varepsilon}_0 - \omega_{k'}} \frac{\partial}{\partial \omega_{k'}} |V_{1/0k'}|^2 \right\} \tag{6.11}$$

This expression has to be compared to (3.13). As may be expected we now have renormalized one-photon processes (see 6.5, 6.6). But the important new feature is that the first bracket no longer vanishes outside resonance. We therefore obtain a finite line breadth. Replacing the δ-function and the principal parts by well known representations as limits of analytic functions one obtains

$$\tilde{\rho}(1_0, 1_k, t)_{t\to\infty} \sim \lambda^2 \frac{1}{(\tilde{\varepsilon}_1 - \tilde{\varepsilon}_0 - \omega_k)^2 + \lambda^4 \tilde{\gamma}_1^2/4} \tag{6.12}$$

This is the usual Lorentz shape. The situation is, however, so simple only because of the Wigner–Weisskopf model used. We cannot go into details here (see Henin, to appear, Physica).

It is interesting to comment about the origin of the broadening. The initial condition refers to the unperturbed excited state. Therefore, in terms of physical states we may expect the other states (including the ground state) also to be excited initially. It is therefore not astonishing that we obtain a broadening due to the rather artificial choice of the initial condition.

Let us now consider resonance scattering. We start with an initial condition similar to (3.16)

$$\tilde{\rho}(1_0, 1_\lambda, t = 0) = 1 \quad \text{all other } \tilde{\rho} = 0 \tag{6.13}$$

In the case of the Wigner–Weisskopf model the physical ground state is identical to the bare one and there is no difference between (3.16) and (6.13). It is necessary to take into account the direct scattering of the photon λ corresponding to (6.7). We shall use a notation similar to (6.5)

$$\theta_{k/k_1} = 2\pi\delta(\omega_k - \omega_{k_1})\sigma(k, k_1) \tag{6.14}$$

One obtains instead of (3.17)–(3.18) the following equations

$$\tilde{\rho}(1_0, 1_\lambda, t)\underset{t\to\infty}{=} 1 - \lambda^2 t[\tilde{\gamma}_{1/0\lambda} - \lambda^2 \sum_k \theta_{\lambda/k}] \qquad (6.15)$$

$$\tilde{\rho}(1_0, 1_k, t)\underset{t\to\infty}{=} \lambda^2 t\left[\frac{\tilde{\gamma}_{1/0\lambda}\tilde{\gamma}_{1/0k}}{\tilde{\gamma}_1} - \lambda^2 \theta_{\lambda/k}\right] \qquad (6.16)$$

As shown by (6.16), the cross section is the *sum of the two* mechanisms discussed in the introduction (§1). In the first term of (6.16) we have the product of the δ-functions (1.9) and in the second the δ-function (1.10). We have, therefore, in this formulation *addition of the various transition probabilities involved*.

Let us first make contact with the *T*-matrix formulation. If the initial photon has an energy different from the renormalized level separation, we may neglect the resonant contribution in (6.16) and obtain for the transition probability $\lambda \to k$ per unit time (see 6.8, 6.14)

$$P_{k/\lambda} = -\lambda^4 \theta_{\lambda/k} = -2\pi\lambda^4\,\delta(\omega_k - \omega_\lambda)\sigma_4(k,\lambda)$$

$$= 2\pi\lambda^4\,\delta(\omega_k - \omega_\lambda)\left[\frac{1}{\tilde{\varepsilon}_1 - \tilde{\varepsilon}_0 - \omega_\lambda}V_{1/0\lambda}V_{1/0k}\right]^2 \qquad (6.17)$$

where we have dropped the *P*-symbol in (6.8) since we are far from resonance. The result (6.17) is identical to that obtained by the *T*-matrix (see §1) for the Wigner–Weisskopf model. This is in fact a general conclusion: whenever the transition process between the initial and the final states can only be achieved through a single sequence we recover the *T*-matrix results directly. As already emphasized in the introduction, the interesting situation for us corresponds to the possibility of different sequences of transitions.

Even near resonance $\theta_{\lambda/k}$ retains a meaning but then we have to keep the *P* symbol in (6.8). This is an essential difference with the Born approximation (6.17) which becomes singular at resonance. We may, therefore, *in all circumstances* add the two contributions in (6.16).

In a similar way as for spontaneous emission the sum can be written as a Lorentz shape (see 6.12).

It may at first seem surprising that the addition of transition probabilities is at all possible because of the well known interference effects of transition amplitudes, which may decrease the total effect when a new 'channel' is opened. A similar phenomenon arises here in the sense that the higher order transition probabilities (such as the λ^4 term in 6.16) are *not necessarily positive*. They are really *excess transition probabilities*. It is only far from resonance that we obtain a perfect square [see for other examples Prigogine, Henin and George (to appear)]. As the result the sum of two sequences may be smaller than the contribution of a single one.

This example, as well as others treated elsewhere, show that the transformation theory summarized in §5 leads to a generalization of Einstein's theory of transition probabilities to higher order in the coupling constant. This is already an interesting result but the most fascinating aspects of this theory are certainly related to the extension of the concept of quantum state and quantum level we have briefly indicated in §5 and which we wish to discuss in more detail first in the simple case of an excited level.

7. Quantum states with finite lifetime

Let us continue to use as an illustration the Wigner–Weisskopf two-level atom. We suppose that at time t_0 the system is described by (see 5.24)

$$\tilde{\rho}(1_1, t_0) = 1 \quad \text{all other } \tilde{\rho} = 0 \tag{7.1}$$

We desire to calculate explicitly the diagonal and off diagonal elements of ρ in the unperturbed representation using (5.25), (5.26), and shall perform this calculation correct to order λ^2. We may then apply (5.8) and obtain the diagonal elements

$$\rho_0(1_1, t_0) = 1 + \lambda^2 \sum_k P \frac{1}{\varepsilon_1 - \varepsilon_0 - \omega_k} \frac{\partial}{\partial \omega_k} |V_{1/0k}|^2 \tag{7.2}$$

$$\rho_0(1_0, 1_k, t_0) = -\lambda^2 P \frac{1}{\varepsilon_1 - \varepsilon_0 - \omega_k} \frac{\partial}{\partial \omega_k} |V_{1/0k}|^2 \tag{7.3}$$

It may be verified that

$$\rho_0(1_1, t_0) + \sum_k \rho_0(1_0, 1_k, t_0) = 1 \tag{7.4}$$

However in the limit of a point particle

$$\begin{cases} \rho_0(1_1, t_0) \to 1 - \lambda^2 \infty \\ \rho_0(1_0, 1_k, t_0) \to +\lambda^2 \infty \end{cases} \tag{7.5}$$

For a consistent quantum mechanical interpretation each diagonal element of the density matrix has to be positive. Therefore such an interpretation seems difficult in the *bare* representation at least for a point particle.

We also have off diagonal elements which in the graphical representation explained in §2 (see Figure 2.1) correspond to diagrams of the type

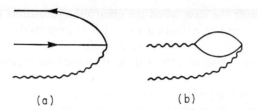

Fig. 7.1. Contributions to off diagonal elements of ρ_0, (a) order λ; (b) order λ^2.

represented in Figure 7.1. The explicit expressions for the off diagonal elements (up to order λ^2) are

$$\langle 1_1 | \rho(t_0) | 1_0 1_k \rangle = -\lambda \left\{ P \frac{1}{\varepsilon_1 - \varepsilon_0 - \omega_k} - i\pi \, \delta(\varepsilon_1 - \varepsilon_0 - \omega_k) \right\} V_{1/0k} \quad (7.6)$$

$$\langle 1_0 1_k | \rho(t_0) | 1_1 \rangle = -\lambda \left\{ P \frac{1}{\varepsilon_1 - \varepsilon_0 - \omega_k} + i\pi \, \delta(\varepsilon_1 - \varepsilon_0 - \omega_k) \right\} V^*_{1/0k} \quad (7.7)$$

$$\langle 1_0 1_k | \rho(t_0) | 1_0 1_{k'} \rangle = \lambda^2 \left\{ P \frac{1}{\varepsilon_1 - \varepsilon_0 - \omega_k} P \frac{1}{\varepsilon_1 - \varepsilon_0 - \omega_{k'}} \right.$$

$$+ \pi^2 \, \delta(\varepsilon_1 - \varepsilon_0 - \omega_k) \, \delta(\omega_k - \omega_{k'}) + i\pi P \frac{1}{\omega_k - \omega_{k'}}$$

$$\left. \times \delta(\varepsilon_1 - \varepsilon_0 - \omega_k) + i\pi P \frac{1}{\omega_k - \omega_{k'}} \delta(\varepsilon_1 - \varepsilon_0 - \omega_{k'}) \right\} V^*_{1/0k} V_{1/0k'} \quad (7.8)$$

As expected we see that to the state (7.1) there corresponds, in the bare representation, a density matrix with both diagonal and off diagonal elements. This matrix is hermitian and we can find a unitary transformation Q which reduces it to the *canonical form* ρ_c

$$\rho_c = Q^+ \rho Q \quad (7.9)$$

with

$$(\rho_c)_{ij} = a_i \delta_{ij} \quad (7.10)$$

Simple calculations show that *if the δ-function contributions in (7.6)–(7.8) are neglected*, then

$$a_i = 1 \delta_{ij} \quad (7.11)$$

Therefore, when dissipation is neglected the specification (7.1) is equivalent to the specification of a wave function. We then have a *pure state*.

This is no longer the case when the complete expressions (7.6)–(7.8) are used. Then a further reduction of (7.10) is impossible. Moreover the interpretation of the different a_i is difficult as they involve products of δ-functions (expressing that the a_i depend on the time necessary to prepare the state).

In the case of the ground state we recover, of course, the usual specification of quantum mechanics (see 5.28). We see very clearly in this simple example that the whole difference lies in the dissipative properties. According to the two possibilities, infinite lifetime or finite lifetime, the specification of the quantum state is radically different*.

Let us consider the quantum mechanical levels. To obtain interesting effects, we have to include terms to order λ^4. The expression for χ to order λ^4 for the Wigner–Weisskopf model is given by

$$\chi = 1 + \frac{\lambda^2}{2}\psi'_2 + \frac{\lambda^4}{2}\psi'_4 + \frac{\lambda^4}{4}[\psi''_2, \psi_2]_+ + \frac{3\lambda^4}{8}\psi'^2_2 + \lambda^4 \chi''_4 \quad (7.12a)$$

where†

$$\chi''_4 = \tfrac{1}{4}\sum_{k_1}\sum_{k_2}\sum_{\varepsilon=\pm 1}\left\{\left[P\frac{1}{\varepsilon_1 - \varepsilon_0 - \omega_{k_1}} P\frac{1}{\varepsilon_1 - \varepsilon_0 - \omega_{k_2}}\right.\right.$$

$$+ \pi^2 \delta(\varepsilon_1 - \varepsilon_0 - \omega_{k_1})\, \delta(\varepsilon_1 - \varepsilon_0 - \omega_{k_2})\bigg]\left(\frac{\partial^2}{\partial \omega_{k_1}^2} - \frac{\partial^2}{\partial \omega_{k_2}^2}\right)$$

$$+ P\frac{1}{\omega_{k_1} - \omega_{k_2}}\left(P\frac{1}{\varepsilon_1 - \varepsilon_0 - \omega_{k_1}} + P\frac{1}{\varepsilon_1 - \varepsilon_0 - \omega_{k_2}}\right)\left(\frac{\partial}{\partial \omega_{k_1}} + \frac{\partial}{\partial \omega_{k_2}}\right)^2\bigg\}$$

$$\times |V_{1/0k_1}|^2 |V_{1/0k_2}|^2 \delta_{N_1,(1+\varepsilon)/2}\, \delta_{N_0,(1-\varepsilon)/2}\left(N_{k_1} + \frac{1-\varepsilon}{2}\right)$$

$$\times \left\{\left(N_{k_2} + \frac{1+\varepsilon}{2}\right)\left[1 - \exp\left\{\varepsilon\left(\frac{\partial}{\partial N_{k_2}} - \frac{\partial}{\partial N_{k_1}}\right)\right\}\right]\right.$$

$$\left.\left.- \varepsilon\left(N_{k_1} + \frac{1+\varepsilon}{2}\right)\left[1 - \exp\left\{-\varepsilon\left(\frac{\partial}{\partial N_1} - \frac{\partial}{\partial N_0} - \frac{\partial}{\partial N_{k_1}}\right)\right\}\right]\right\}\right] \quad (7.12b)$$

---

* To avoid misunderstanding let us stress that this difference is *not* related to a difference between a bound state and an elementary particle.

† We have adopted the 'natural order of integration' introduced in Henin, Prigogine, George, Mayné (1966) to give a well-defined meaning to the product of principal parts. Notice that, due to the Poincaré-Bertrand theorem, equivalent forms of χ''_4 can be given in which no delta functions appear.

Using (5.22) and (5.24) we obtain

$$\tilde{\varepsilon}_1 = \varepsilon_1 + \lambda^2 \sum_k P \frac{1}{\varepsilon_1 - \varepsilon_0 - \omega_k} |V_{1/0k}|^2$$
$$+ \lambda^4 \sum_{kk'} \left\{ P \frac{1}{\varepsilon_1 - \varepsilon_0 - \omega_k} P \frac{1}{\varepsilon_1 - \varepsilon_0 - \omega_{k'}} \right.$$
$$\left. - \frac{\pi^2}{2} \delta(\varepsilon_1 - \varepsilon_0 - \omega_k) \delta(\varepsilon_1 - \varepsilon_0 - \omega_{k'}) \right\} \frac{\partial}{\partial \omega_k} |V_{1/0k}|^2 |V_{1/0k'}|^2 \quad (7.13)$$

This expression corresponds to the case in which no real photons are present ($N_k = 0$). The term in λ^2 is well known and may be obtained by other approaches such as those based on Green's functions (Goldberger and Watson (1964)).

The term in λ^4 is different from what the Green's function approach would give (see also §10). This is not astonishing as in this term we have already the 'dissipative' contributions containing δ-functions. If they could be neglected we would come back to the usual result. The presence of such types of terms is, of course, extremely interesting as they indicate the influence of lifetime on the energy levels.

However it is clear that an effect such as the Lamb shift measured by resonance scattering can be calculated through conventional S-matrix theory as no unstable asymptotic states are involved. Whether the consideration of redefined 'real' energy states for the excited atom would simplify the calculation is another matter which has to be tested.

We may now, using (7.3) express the thermodynamic properties of an atom (for example in equilibrium with a black body) in terms of *particle states* (see 1.8). As a result we see that the specific heat would be given (for T sufficiently low) exactly by the same formula as in the absence of coupling with the transverse electromagnetic field but with the redefined level (7.13).

We may even say that this level has been chosen in such a way as to make this result identical to what a direct calculation based on the canonical distribution (1.6) would provide.

This leads us to consider a little more closely the relation between our new concept of quantum states and thermodynamic considerations based on the entropy.

8. Physical particles and entropy

We have obtained a description of the time evolution of the system expressed by (5.11). By a suitable choice of our dressing operator (our Boltzmann representation, see §5) this evolution is represented in terms of real energy conserving scattering processes. This implies a modification of the energy levels and of the interaction energy. In other words we have energy (or mass) and coupling renormalization. For non dissipative situations (atom in its ground state, stable V particle in the Lee model, see §10) the energy renormalization is identical to that of usual field theory.

Also, at least in the frame of the perturbation expansion used we obtain for the Lee model (with stable V particles) the usual charge renormalization. As all contributions to the collision operator φ are now energy conserving, no virtual processes now appear in the evolution of $\tilde{\rho}$. It is therefore natural in the frame of our theory to consider $\tilde{\rho}$ as the distribution function of the physical (interacting) particles.

For free fields the definition of particles is, of course, trivial. This is however no longer the case for *interacting fields*. If we could by some canonical transformation reduce the Hamiltonian to a sum of independent terms there would be no problem, but that is seemingly out of the question. Therefore we have to invoke other considerations to introduce the particles associated with interacting fields. The idea we have followed is to make use of the transformation theory of §5 to obtain a representation of the distribution function in which the entropy both at equilibrium and out of equilibrium may be considered as *purely combinatorial**.

To each pure state such as (7.1) there corresponds a zero entropy. The value of the entropy is related to, as in quantum statistics of perfect gases, the permutations between the different possible quantum states.

One could say that we use the classical argument about entropy in its reversed form: one proves that particles, when weakly coupled, have a purely combinatorial Boltzmann entropy. We put the entropy into the combinatorial form and conclude that the particles are then well defined physical entities!

This introduction of thermodynamic considerations is not so surprising as it may seem at first. After all, the description in terms of physical particles is already a contracted description in comparison with the initial field description. We shall come back to the thermodynamic implications of our approach in §11.

What, then, is the difference between our renormalized theory and the usual field theoretical approach? Probably the most striking is that in our approach *no renormalized field can be defined*.

The distribution function $\tilde{\rho}$ can no longer be factorized into a product

---

* It is amusing to notice the analogy of this argument with Planck's thermodynamic derivation of the radiation formula which is also based on the combinatorial expression of entropy (Planck, 1901, see also Rosenfeld, 1958). See also the footnote on p. 202.

of two wave functions. One can say that *the wave (or field) concept has disappeared as a dynamical* concept. It is true that all states corresponding to an infinite life time are still described by renormalized wave functions*, but we have all the states with finite life time for which this is no longer true! The concept of wave function renormalization survives in our approach in a much more *restricted* sense: as the matrix element of the operator χ connecting the bare ground state to the physical ground state (see §9).

To make these remarks more concrete let us briefly consider the two standard field theoretical models: the neutral, scalar field with fixed c-number point sources and the Lee model.

9. Neutral scalar field with fixed sources

We shall use the notations of Barton (1963) and quote only a few results to make a comparison in an especially simple case between the usual field theoretical concepts and our method.

The starting point is the Hamiltonian (V is here the volume)

$$H = \sum_k \omega_k a_k^+ a_k + \lambda V^{-1/2} \sum_{k,i} \frac{\mu(\omega_k)}{(2\omega_k)^{1/2}} \left\{ a_k e^{-ikx_i} + a_k^+ e^{ikx_i} \right\} \quad (9.1)$$

As is well known there exists a unitary transformation U (given for example in Barton, 1963) which transforms (9.1) into

$$H = \sum_k \omega_k b_k^+ b_k - \lambda^2 V^{-1} \sum_{kij} \frac{\mu^2(\omega_k)}{2\omega_k^2} e^{ik(x_i - x_j)} \quad (9.2)$$

In this representation we have free mesons plus the potential energy of the sources.

Of course, our theory is not necessary for this situation in which no dissipative processes occur at all. However, we may calculate the dressing operator χ to order λ^4 using formulae such as (7.12). The result may be written in the exponential form

$$\chi = \exp\left[\frac{\lambda^2}{2V} \sum_{\substack{k \\ ij}} \frac{\mu^2(\omega_k)}{\omega_k^3} \right.$$

$$\left. \times \exp\left\{ ik(x_i - x_j) \right\} \left\{ n_k (1 - e^{-(\partial/\partial n_k)}) + (n_k + 1)(1 - e^{+(\partial/\partial n_k)}) \right\} \right] \quad (9.3)$$

---

* In other words the concept of a wave function retains its entire validity as an eigenfunction of a non-dissipative state.

We then find that the collision operator (5.7) vanishes identically

$$\varphi = 0 \qquad (9.4)$$

This, of course, expresses the non-dissipativity of the model. Moreover the Hamiltonian H_R of our theory, as given by (5.22), becomes identical to (9.2). This is quite gratifying as in this case the transition from (9.1) to (9.2) is precisely equivalent to a transformation to a representation in which the bosons no longer interact. Therefore, they will have an entropy expressed in terms of the usual Bose–Einstein statistics.

We also have a direct relationship between the operator χ and the unitary operator U which leads from (9.1) to (9.2). Let us write (5.10) more explicitly in the form

$$\tilde{\rho}_n = \sum_{n'} \langle n | \chi^{-1} | n' \rangle \rho_{n'}^0 \qquad (9.5)$$

Using (9.3) we see that

$$\langle 0 | \chi^{-1} | 0 \rangle = \exp\left[-\frac{\lambda^2}{2V} \sum_{kij} \frac{\mu^2(\omega_k)}{\omega_k^3} e^{ik(x_i - x_j)} \right]$$

$$= \langle 0 | U | 0 \rangle \qquad (9.6)$$

where $\langle 0 | U | 0 \rangle$ is the projection of the physical vacuum (of conventional field theory) on the bare vacuum [Barton (1963), formula (13.26)]. In the limit of point particles it is well known that all matrix elements of U vanish in a given representation. The same is true for χ^{-1}. In other words our theory appears as a natural extension of 'improper' or 'inequivalent' representations to dissipative situations for which χ can still be defined (while an unitary transformation U which would reduce the initial Hamiltonian to a sum of independent parts is then not likely to exist).

10. Lee model—ghost states

Let us now consider the Lee model [see for example Barton (1963), Schweber, Bethe and de Hoffmann (1956)].

$$V \rightleftharpoons N + \theta \qquad (10.1)$$

This model already involves scattering processes and is therefore of great interest to test our approach. We have derived the dressing operator χ to order λ^4 and we may therefore obtain the Hamiltonian H_R and the collision operator φ up to this order (see 5.7 and 5.22). From these expressions we deduce the energy per particle (see 5.24) and the charge renormalization. Our method is valid for all sectors. For the sector $B = 1$, $Q = 0$* we recover for *stable* V particles the well known results of field theory.

---

* We use again the notation of Barton (1963), $B = n_V + n_N$, $Q = n_\theta - n_N$.

However, for *unstable* V particles we obtain different results. The ground state energy of a V particle of momentum p_0 is found to be in the sector $B = 1$, $Q = 0$ (δ^k is a Kronecker delta function, Ω is the volume)

$$\tilde{\varepsilon}_{V_{p_0}} = \varepsilon_{V_{p_0}} - \lambda^2 \sum_{kq} \frac{\delta^K(p_0 - q - k)}{(-\varepsilon_{V_{p_0}} + \varepsilon_{N_q} + \omega_k)_P} \frac{1}{2\omega_k \Omega}$$

$$+ \lambda^4 \sum_{kq} \sum_{k'q'} \delta^K(p_0 - q - k) \delta^K(p_0 - q' - k') \left\{ -\frac{\pi^2}{2} \delta(-\varepsilon_{V_{p_0}} + \varepsilon_{N_q} + \omega_k) \right.$$

$$\left. \times \delta(-\varepsilon_{V_{p_0}} + \varepsilon_{N_{q'}} + \omega_{k'}) + \frac{1}{(-\varepsilon_{V_{p_0}} + \varepsilon_{N_q} + \omega_k)_P} \frac{1}{(-\varepsilon_{V_{p_0}} + \varepsilon_{N_{q'}} + \omega_{k'})_P} \right\}$$

$$\times \frac{\partial}{\partial(-\varepsilon_{V_{p_0}} + \varepsilon_{N_q} + \omega_k)} \frac{1}{4\omega_k \omega_{k'} \Omega^2} \quad (10.2)$$

In the case of a stable V particle the δ-functions give vanishing contributions and (10.2) simply becomes the solution correct to order λ^4 of the usual eigenvalue equation.

$$\tilde{\varepsilon}_{V_{p_0}} = \varepsilon_{V_{p_0}} - \lambda^2 \frac{1}{\Omega} \sum_{kq} \frac{\delta^K(p_0 - k - q)}{2\omega_k} \frac{1}{-\tilde{\varepsilon}_{V_{p_0}} + \varepsilon_{N_q} + \omega_k} \quad (10.3)$$

For the case of an unstable particle Glaser & Källen (1956) have formulated an eigenvalue problem (see also Levy, 1959) which leads to

$$m_V = m_V^0 - \frac{\lambda^2}{\Omega} \sum_k \frac{1}{(-m_V^0 + m_N^0 + \omega_k)_P} \frac{1}{2\omega_k}$$

$$+ \frac{\lambda^4}{\Omega^2} \sum_k \sum_{k'} \left\{ \frac{1}{(-m_V^0 + m_N^0 + \omega_k)_P (-m_V^0 + m_N^0 + \omega_{k'})_P} \right.$$

$$\left. - \pi^2 \delta(-m_V^0 + m_N^0 + \omega_k) \delta(-m_V^0 + m_N^0 + \omega_{k'}) \right\} \frac{\partial}{\partial \omega_k} \frac{1}{2\omega_k \cdot 2\omega_{k'}} \quad (10.4)$$

This result agrees with (10.2) *except* in the coefficient involving the product of δ-functions where we find a supplementary factor $\frac{1}{2}$*. This difference appears at order λ^4. We want to state explicitly that we do not consider Glaser & Källen's result as 'wrong' and ours as 'exact'. We just don't calculate the same thing. However what we claim is that it is in terms of our expression that thermodynamic quantities (such as the partition functions) as well as the cross sections can be expressed in terms of *one-particle states*.

* This factor has been checked by different independent calculations. It appears also in (7.13).

The difference between our approach and the usual field theoretical one appears also in a different context: *We have no 'ghost' states.* To understand this important point let us make the following two remarks: (1) The distribution function $\tilde{\rho}$ of the physical particles has been chosen in such a way that the entropy is given by the Boltzmann expression (5.19) and that moreover this expression coincides at equilibrium with the canonical entropy (which is independent of any definition of the physical particle states). If ghost states were to exist in our theory they would have to give a *vanishing contribution to all long time macroscopic properties which may be derived* from the partition function. This is very unlikely.

(2) We have already stressed in §8 that the field as a dynamical concept has disappeared in our theory. There exists nothing like *renormalized* field operators. Renormalized propagators cannot be defined in our theory and therefore the very source of the ghost states (which are related to the poles of the *renormalized* propagators) has disappeared.

Therefore it seems that we obtain a consistent dynamical description involving *physical* particles. However this description is 'out' of the Hilbert space as all states corresponding to a finite lifetime are defined outside this space. More generally the fundamental Lehmann, Symanzyk and Zimmermann theorem (1955) (see also Barton, 1963) leads to no difficulty as renormalized vacuum expectation functions cannot be defined in our formulation.

11. Quantum states and entropy—irreversibility and dressing

Let us make somewhat more explicit the general considerations of §8. We have already stressed the importance of the entropy concept to define physical particles. The expression of entropy (or of the $\mathcal{H}$-quantity, with $S = -k\mathcal{H}$ we used was given in §5

$$\mathcal{H} = \sum_N \tilde{\rho}(N) \log \tilde{\rho}(N) \tag{11.1}$$

This expression is constructed to give both an $\mathcal{H}$-theorem (increase of entropy due to irreversible processes) and the correct entropy at equilibrium. Let us recall that there exists a second statistical expression for entropy, the so called Gibbs entropy (see R. Tolman, 1938)

$$\mathcal{H}_{\text{gibbs}} = \text{tr}\, \rho \log \rho \tag{11.2}$$

For *equilibrium* both expressions give identical results. But out of equilibrium, (11.2) cannot be valid as it remains constant in time. This situation has been recently analyzed in great detail [Prigogine and Henin (1967), Prigogine, Henin and George (1967), Nicolis (1967), Philippot and

Walgraef (1967)]. We even know how to extract from (11.2) the 'thermodynamic' part (11.1). Of course, the difference between (11.1) and (11.2) vanishes at equilibrium. We may present the following two statements.

(1) There are pure states (that is states characterized by a well defined wave function in Hilbert space) which give to the entropy (11.1) a *non-vanishing* value. For example to the unperturbed excited atomic state (6.10) there corresponds, using (5.10), a whole set of physical states $\tilde{\rho}(N)$ and therefore a non vanishing entropy. (This of course would be out of the question if the Gibbs formula (11.2) were applicable);

(2) There are statistical mixtures with a *vanishing entropy*. This is the inverse of statement (1). This is a direct consequence of the example studied in §7. To the physical state (7.1) corresponds a vanishing entropy (11.1) but in the bare representation it is represented, as we have seen, by a statistical matrix which cannot be reduced to a pure state.

We see, therefore, that the distinction between pure and mixed states, so basic, for example, in von Neumann's work (1932) (see also London and Bauer, 1939) does *not* correspond to a distinction between vanishing and non-vanishing entropy. On the contrary each state $\tilde{\rho}$ taken separately corresponds to a state of zero entropy (or maximum information).

As a consequence we may introduce a 'phase space' formed by cells representing each a possible state $\tilde{\rho}(N)$ (see Figure 11.1). For transitions between two non-dissipative states, both represented by well defined wave functions, the S-matrix theory remains valid. However, we may also consider more general transitions (even for transitions between two non-dissipative states we obtain a different *description*, see for an example §6,

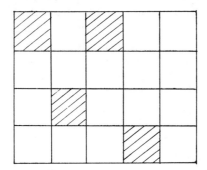

Fig. 11.1. Schematic representation of the phase space.

▨ non dissipative quantum states $\tilde{\rho}$ corresponding to wave functions in Hilbert space.

☐ dissipative quantum states $\tilde{\rho}$ not in Hilbert space.

but the result coincides to all orders to which we are able to make the calculations with the S-matrix results).

It should also be noticed that quantum states themselves appear through an asymptotic procedure which is based on the existence of a dissipative mechanism (see 2.13).

More precisely we can now analyze the transition from an initial bare state to physical states and verify using (11.1) that the *dressing process in dissipative systems is an irreversible process involving an increase of entropy*. It is precisely because it can take into account this 'thermodynamic' aspect of renormalization in dissipative systems such as coupled fields, that our theory goes beyond the usual purely mechanical approach.

12. General conclusions*

We are well aware of the limitations of the method we have used and which relies heavily on perturbation theory. It may however be hoped that a non perturbative approach if at all possible would retain the qualitative features we have discussed in this report.

The main point is probably the appearance of a kind of duality (or complementarity) in the physical description: we could use the bare interacting fields, but we would then have to give up the particle description† (no distinction between real and virtual processes can be made) or we use a description involving physical particles but then the field as a dynamical concept is eliminated. (However as repeatedly emphasized it retains its strict validity when associated to stable states).

This surprising situation results from the infinite number of degrees of freedom involved in the system we study. Certainly we may, for $t = 0$, prescribe the fields at all points. But immediately afterwards, for $t > 0$, we must expect irreversible rearrangements related to the dressing of the particles (see §11). The physical particles emerge as the result of an asymptotic procedure. The 'real' initial condition is lost (see 5.2) and with it the exact wave function of the system. From this point on we deal only with a statistical ensemble.

It seems to us that our approach links together various elements, some of which have been discussed in the literature. The difficulty of extending field theories to situations other than free field is well recognized and associated with the Haag theorem (Haag, 1955). The possible rôle of improper unitary transformations has been repeatedly stressed (see Barton, 1963). While we start from a field theory based on a canonical

---

* See also §§8, 11, and the footnote on p. 202.

† This would however lead to great difficulties as in no experiment is the field observed directly.

formalism we obtain, once the asymptotic elements are introduced (such as continuous spectrum, 'large' times) a new mathematical structure associated with a particle description. However, there still exists a phase space (see §11) and S-matrix theory remains valid for transitions for which asymptotic states are defined.

Therefore in this sense *S-matrix theory appears indeed as more general than field theory* (see e.g. Chew, 1961). In other words a consistent dynamical description of the physical situation involving interacting fields is not necessarily a field theory in the conventional sense. Some elements of the canonical formalism are lost and the statistical aspects of description are reinforced. We are even farther away from the classical deterministic description of nature*. However, there exists in this description a 'phase-space' (see §11) formed both by the stable and unstable states of the system. A further investigation of the transitions between these states as well as the extension of the concepts involved appears to us as our next task.

Acknowledgements

The results presented in this report have been obtained during the last two years by our group in Brussels. A specially important part has been played by Dr. Cl. George, Dr. F. Henin and Dr. Résibois. But I am also indebted to MM. Mayné, Mandel, De Haan for calculations and discussions. I also wish to thank Professor Géhéniau for his interest and encouragement.

Various aspects of this work have been sponsored by the following institutions to which I wish to extend my appreciation for continuous support:
Fonds National Belge de la Recherche Fondamentale Collective.
Dupont de Nemours International (*Geneva*).
The Air Force Office of Scientific Research through the European Office of Aerospace Research, *OAR*, United States Air Force under contract AF EOAR67-25.

References

Balescu, R. (1963), *Statistical Mechanics of Charged Particles* (Interscience, New York).
Balescu, R. (1968), *Physica*, **32**, 98.
Balian, R. and de Dominicis, C. (1960), *Compt. rend.*, **250**, 3285, 4111.
Balian, R. and de Dominicis, C. (1960), *Nucl. Phys.*, **16**, 502.
Barton, G. (1963). *Introduction to Advanced Field Theory*, Interscience, New York and London.

* This conclusion has been anticipated by Rosenfeld (1957).

Baus, M. (1966), *Phys. Fluids*, **9**, 1427.
Baus, M. (1967), *Acad. Roy. Belg.*, *Bull. Classe Sci.*, **53**, 1291; **53**, 1332; **53**, 1352.
Chapell, W. R., Britten W. E. and Glass, S. J. (1965), *Nuovo Cimento*, **38**, 1168.
Chew, G. (1961), see *XIIth Solvay Conference on Quantum Field Theory*, p. 173, Interscience, New York, London.
Einstein, A. (1917), *Physikalische Zeitschrift*, **18**, 121.
George, C. (1964), *Physica*, **30**, 1513.
George, C. (1967), *Physica* **37**, 182.
George, C. (to appear).
Glaser, V., and Källen, G. (1956), *Nucl. Phys.*, **2**, 706.
Goldberger, M. L., and Watson, K. M. (1964), *Collision Theory*, Wiley, New York.
Grosjean, K. (1959), *Théorie Formelle du Scattering*, Institut Interuniversitaire des Sciences Nucléaires.
Haag, R. (1955), *Mat. Fys. Ned.*, **29**, 12.
Heitler, W. (1954), *The Quantum Theory of Radiation*, 3rd ed., Clarendon Press, Oxford.
Henin, F. (1963), *Physica*, **29**, 1233.
Henin, F., Prigogine, I., George, C. and Mayné, F. (1966), *Physica*, **32**, 1828.
Kroll, N. M. (1965), in *Quantum Optics and Electronics*, Summer School at Les Houches 1964, ed. De Witt, C., Blandin A. and Cohen C. -Tannoudji, (Gordon & Breach).
Lehmann, H., Symanzik, K. and Zimmermann, W. (1955), *Nuovo Cimento*, **2**, 425.
Lévy, M. (1959), *Nuovo Cimento*, **13**, 115.
London, F. and Bauer, E. (1939), *La théorie de l'observation en Mécanique Quantique*, Herman, Paris.
Mangeney, A. (1961), *Physica*, **30**, 461.
Nicolis, G., *J. Chem. Phys.*, **46**, 702 (1966).
Nozières, P. (1963), *Le Problème à N corps*, Dunod, Paris.
Philippot, J., and Walgraef, D. (1966), *Physica*, **32**, 1283.
Philippot, J., and Walgraef, D. (1967), *Report at the IUPAP International Conference on Statistical Mechanics and Thermodynamics*, ed. T. A. Bark, Benjamin, New York.
Philippot, J., and Walgraef, D. (1967), *Physica*, **33**, 671.
Planck, M. (1901), *Ann. Physik*, **4**, 553.
Prigogine, I. (1962), *Non Equilibrium Statistical Mechanics* (Interscience, New York).
Prigogine, I. (1967), *Proc. of the International School on Non Linear Math. & Phys.*, Munich 1966 (Springer Verlag, to appear).
Prigogine, I. and George, C., (to appear).
Prigogine, I. and Henin, F. (1960), *J. Math. Phys.*, **1**, 349.
Prigogine, I. and Henin, F. (1967), *Report at the IUPAP International Conference on Statistical Mechanics and Thermodynamics*, ed. T. A. Bark, Benjamin, New York.
Prigogine, I., Henin, F., and George, C. (1966), *Physica*, **32**, 1873.
Prigogine, I., Henin, F., and George, C. (1968), monograph in preparation.
Prigogine, I. and Leaf, B. (1959), *Physica*, **25**, 1067.

Prigogine, I. and Severne, G. (1966), *Physica*, **32**, 1376.
Prigogine, I., and Severne, G. (1967), *Colloquium on the n-body gravitation problem*, Paris (to appear).
Résibois, P. (1963), *J. Math. Phys.*, **4**, 166.
Résibois, P. (1963a), *Physica*, **29**, 721.
Résibois, P. (1965), *Physica*, **31**, 645.
Résibois, P. (1965a), *Phys. Rev.*, **138**, B281.
Résibois, P. (1966), in '*Physics of Many Particle Systems: Methods and Problems*', ed. Meeron, E. (Gordon and Breach, New York).
Résibois, P. (1966a), *Physica*, **32**, 1473.
Résibois, P. and De Leener, M. (1966) *Phys. Rev.*, **152**, 305.
Rosenfeld, L. (1957), in *Observation & Interpretation*, p. 47, ed. S. Körner, London.
Rosenfeld L. (1958), Max Planck Festschrift, 203.
Schieve, W. C. (1967), *Physica*, **34**, 81.
Schweber, S., Bethe, H., and de Hoffman, F. (1956) *Mesons and Fields*, Vol. I (Row Perterson & Company, Evanston, Illinois).
Schweber, S. (1961), *Introduction to Relativistic Quantum Field Theory*, Row Peterson & Co., New York.
Severne, G. (1965), *Physica*, **31**, 877.
Tolman, R. (1938), *The Principles of Statistical Mechanics*, Oxford, Univ. Press.
von Neumann, J. (1932), *Mathematische Grundlagen der Quantenmechanik*, Springer Verlag, Berlin.
Watabe M. and Dagonnier R. (1966), *Phys. Rev.*, **143**, 110.
Zwanzig, R. (1960), *J. Chem. Phys.*, **33**, 1338.
Zwanig, R., *Lectures Theor. Phys.*, (Boulder, vol. III, p. 106).
Zwanzig, R. (1964), *Physica*, **30**, 1109.

Discussion on the communication of I. Prigogine

G. Källén. I agree with most of what you have said—at least to the extent that I have understood it. In particular, I am not at all surprised by the fact that alternative definitions of the mass give results which differ by amounts of the order of magnitude of the inverse life time.

However, I should like to comment that most contributions from ghosts are non analytic in λ and, therefore, never seen in a perturbation theory expansion in λ. They will probably appear again if you can treat the problem without perturbation theory—at least to the extent that they are there.

Finally, I have a question. Most people believe that the concept of an unstable particle will become rather uncertain in the limit when the lifetime is very short (strong interactions). Presumably, you do not see this because you are using perturbation theory. Do you have any idea about how the unstable particle is going to disappear in the limit $\tau \to 0$ in your formalism?

I. Prigogine. As we use perturbation theory (up to order λ_0^4) we obtain for the charge renormalization in the Lee model

$$\lambda^2 = \lambda_0^2(1 - \lambda_0^2 L) \tag{1}$$

instead of

$$\lambda^2 = \lambda_0^2(1 + \lambda_0^2 L)^{-1} \tag{2}$$

If we now adjust in (1) λ^2 to a given 'experimental' value, this may imply for sufficiently large values of L a complex value for λ_0, that is a non hermitian Hamiltonian. Still, even in this case, we do not see in our representation $\tilde{\rho}$ of the physical particles, any trace of ghost states. If such states do exist, they would have for some reason to give no contribution to the entropy which is already taken into account entirely by the 'physical' states. Of course, as we cannot define in our formalism renormalized field operators, we cannot use the completeness condition in Hilbert space (in conjunction with the V.E.V. of renormalized propagators) to test the existence of ghost states.

For all those questions it would indeed be important to get rid of the perturbation theory. As long as this is not done, I cannot discuss the limit of short lifetimes.*

G. F. Chew. It seems to me that there is no ambiguity about the definition of an unstable particle. The macroscopic space–time interpretation of the S-matrix is unambiguous, as shown in great detail recently by Iagolnitzer

_____
* See the footnote, p. 202.

and the physical phenomenon corresponding to a resonance reflects the presence of a complex pole in the S-matrix. The 'mass' of the resonance can be unambiguously defined as the pole position.

M. Lévy. I would like to say that I agree generally with what Källén has said about the uncertainty of order λ^4 on the mass of the unstable particle, which is a direct consequence of the uncertainty principle. However, I do believe that the existence of a pole in the other Riemann sheets for the scattering amplitude provides an unambiguous definition for the mass and lifetime of an unstable particle.

On the other hand, I would like to ask you two questions:

(1) Might the possible existence of ghost states in the Lee model reflect itself in the non-positive definition of your density matrix $\tilde{\rho}$?

(2) Does your method, which I find an interesting and natural method to treat the evolution of unstable systems, enable you to predict the decay law (or the evolution in time) of an unstable system? This is a problem which has worried many of us and which does not seem completely solved at present.

I. Prigogine. In our representation $\tilde{\rho}$ there is a sharp value for the mass of unstable particle. This is not in conflict with the uncertainty principle as, when we go back to the bare representation ρ, we obtain a spread in energy precisely of the order required by the uncertainty principle.

We don't question the existence of a pole in the second Riemann sheet with well defined real and imaginary parts. The question is, what is their physical meaning? (See §10 of my paper.)

My answer to your two questions is the following:

(a) The diagonal density matrix $\tilde{\rho}$ can be considered as positive. The usual difficulties appear however if we want to go back to the bare representation ρ (see §7). There a consistent quantum mechanical interpretation seems impossible for a point particle. But this is not a real difficulty as only the physical representation $\tilde{\rho}$ is assumed to be accessible experimentally.

(b) In the frame of the perturbation method used, the evolution of an unstable system is given by our kinetic equation supplemented by a suitable initial condition.

G. F. Chew. The macroscopic space–time content of the S-matrix is complete, including the capacity to describe the formation of a resonance as well as its decay.

D. Ruelle. If one wants to be pedantic there is a little problem in defining the mass of an unstable particle as the position of a pole in the S-matrix, namely to prove that this pole has the same position for different channels.

G. F. Chew. The analytic continuation of the unitary condition leads immediately to the requirement that a pole should appear at the same point for all channels with which it communicates.

E. C. G. Sudarshan. The demonstration that normalizable quantum mechanical states could not always be factorized into (pure state) wave functions is really interesting. Sometime ago Falkoff and Premanand (*Phys. Rev.*, in press) studied the question of equilibrium thermodynamics of strongly coupled system: as a special model they took matter (collection of two-level atoms) and radiation. They found—to their amazement—that the equilibrium entropy and partition function did not at all resemble that for a non interacting matter-radiation system. They used the Boltzmann equation in terms of the Master Equation and used all the exact conservation laws.

Is it likely that this situation also corresponds to this non-factorizability?

I. Prigogine. I don't know the paper you mention but it is very likely that their difficulties could indeed have arisen from the use of a 'non-physical' representation for which a particle interpretation is difficult. At least that was indeed the difficulty in the interesting work by Mangeney on plasmas interacting with an electromagnetic field.

E. P. Wigner. Could we have some more detailed explanation of the equation which gives the time derivative of the diagonal part of the density matrix in its original form? Evidently, such an equation implies some assumption concerning the non-diagonal elements of the density matrix which permitted their elimination from the equation. It would be good to know what this assumption is.

It would also be good to understand the meaning of the term *dissipation* in the present, field-theoretic context; as it was pointed out, the final equations, for the transformed density matrix, will represent a good approximation only if there is some kind of dominance of the dissipative process. A more detailed picture of this process would be helpful in visualizing the conditions of such dominance.

I. Prigogine. I thank Professor Wigner for his very interesting question which gives me the opportunity to clarify some of the important points in my talk.

To begin with, I should mention that I have restricted myself to the description of homogeneous systems, i.e. systems such that all off diagonal elements which are different from zero at $t = 0$ can be connected to the diagonal elements by means of a certain number of application of the operator δL.

The first step in our method mainly consists in a rewriting of the original von Neumann–Liouville equation. To do this, we in fact work on the formal solution of the equation. First we consider the diagonal elements

ρ_0 of the density matrix and recognize the fact that their evolution is due to two mechanisms: collisions and influence of the initial correlations; this gives rise to the two terms in Eqn. (2.10). Then we consider the evolution of correlations (off-diagonal elements); there again we split them into two parts: one, ρ'_ν, which satisfies an equation similar to that of the diagonal elements and describes the scattering of the initial correlations. The second is ρ''_ν which expresses the creation of fresh correlations from the vacuum of correlations and states of a lower degree of correlation. So far all equations are exact. The next step is to ask whether, when we take into account the fact that we have a large number of degrees of freedom, we can obtain a simpler description for long times.

The simplest situation, which is the only one I have considered in this paper, corresponds to what we call the kinetic limit. This is a situation which arises when, among all characteristic time scales which can be defined, there is one, the relaxation time, which is much longer than all the others. In that case, it can be shown on very simple models that for times of the order of the relaxation time, the system has forgotten its initial condition; more precisely, all that remains from the initial condition is taken into account in a redefinition of the initial condition (post initial condition) but the initial correlations no longer influence the time evolution of the system. This is reflected in the transition from the von Neumann equation to the kinetic equations and corresponds to the following properties:

(a) the destruction term in the equation for the diagonal elements vanishes.

(b) the part ρ'_ν of the correlations vanishes.

(c) the part ρ''_ν of the correlations can be expressed in terms of ρ_0 alone.

It should be emphasized that, whereas the original von Neumann equation derives from a Hamiltonian, this is no longer the case for the kinetic equation (simple examples of kinetic equations are of course the Boltzmann equation, the Pauli equation ...). Also, we should mention that if we think in terms of the density matrix for the *complete* system, the kinetic equations are never valid; the only possible description is the von Neumann equation. The kinetic description is valid only if the asymptotic density matrix is used to compute the evolution of observable quantities which depend only on a finite number of degrees of freedom (intensive properties) and involve only correlations of microscopic range. The latter condition means that we shall never consider the correlation individually but rather wave packets. Condition (b) then really means that the spreading of the wave packet is sufficient to allow us to neglect the contribution of ρ'_ν to the asymptotic properties.

Although these assumptions might seem quite plausible when dealing for instance with dilute gases interacting through short range forces, one may wonder whether they are not too drastic in a field theoretical context. Let us for instance discuss the interaction between light and atoms and let us consider two kinds of rather different situations.

First we consider a problem at zero temperature: the spontaneous decay of an excited level. We restrict ourselves to a two level atom (ground state $|0\rangle$, excited state $|1\rangle$) and assume that the only non-vanishing matrix element of the interaction is $V_{1/0k}$, i.e. describes absorption with a transition from the ground state to the excited state or the inverse transition with emission of a photon. For such a system we can define 4 time scales: (a) the collision time $\tau_c = \omega_0^{-1}$ where ω_0 is the cut-off frequency (linked with the Compton wave length for instance), (b) the Bohr time $\tau_B = \hbar/(\varepsilon_1 - \varepsilon_0)$ linked with the Bohr frequency, (c) the Lamb shift time $\tau_{LS} = \hbar(\Delta E_{LS})^{-1}$, (d) the relaxation time $\tau_R = \gamma^{-1}$ where γ is the inverse lifetime of the excited state. In the above model, within the dipolar approximation

$$|V_{1/0k}|^2 = \lambda^2 V^2 \frac{1}{k} \tag{1}$$

we have

$$\tau_{LS}^{-1} \sim \lambda^2 V^2 (\varepsilon_1 - \varepsilon_0) \ln \frac{\hbar\omega_0}{\varepsilon_1 - \varepsilon_0} \tag{2}$$

$$\tau_R^{-1} \sim \lambda^2 V^2 (\varepsilon_1 - \varepsilon_0). \tag{3}$$

With $\hbar\omega_0 \gg \varepsilon_1 - \varepsilon_0$, and $\lambda^2 V^2 \ll 1$ we obviously have:

$$\tau_C \ll \tau_B \ll \tau_{LS} \ll \tau_R. \tag{4}$$

The non-markovian corrections which are kept in the solution of the kinetic equation can be shown to be of the form:

$$\alpha \frac{\tau_B}{\tau_{LS}} + \beta \frac{\tau_B}{\tau_R} \tag{5}$$

where α and β are numerical constants of order 1.

One can also verify in that case that the terms which have been neglected in the transition from the Liouville equation to the kinetic equation are of two kinds: first, we have contributions which depend on τ_c/τ_R [with a cut-off function of the form $\omega_0^4/(\omega^2 + \omega_0^2)^2$ for instance; these contributions are proportional to $\exp(-t/\tau_C)$]. In any case, such contributions are negligible in the limit $\tau_c \to 0$ (such a limit may always be taken for non-divergent contributions). The other contributions, which are related to the fact that the operator $\psi(z)$ has branch point singularities because of the finite lower bound for the photon spectrum, are proportional to $(\tau_B/\tau_R)(\tau_B/t)^n$ where n is an integer ≥ 2. The kinetic equation does not take such contributions (the often discussed corrections to exponential decay) into account. With the inequalities (4), this means that the kinetic equation is valid if we neglect all phenomena characterized by the smallness parameter τ_B/τ_R. (We notice that this does not imply a weak coupling assumption; indeed (5) shows that besides contributions in τ_B/τ_R which should be neglected here the non-markovian corrections contain terms proportional to τ_B/τ_{LS}.) This example shows us that we may expect, in problems at zero temperature, that the kinetic equation will provide a reasonable description when we are dealing with levels with a long lifetime (as for example the $2S$ level in the H atom); however when the lifetime is not the longest characteristic time scale (as for instance for the $2P$ levels in the H atom where τ_{LS} is of the same order as τ_R), the kinetic equation will give us correctly the exponential part of the decay (i.e. the evolution for $t \simeq \tau_R$ as well as the state of the system for $t \to \infty$ (line shape) but corrections should be added when $\tau_R \ll t \ll \infty$.

In the case of spontaneous emission, we start with an initial condition where the density matrix is diagonal in the unperturbed Hamiltonian representation. In that case, there are certainly no problems with the destruction term (2.10) and with the part ρ'_ν of the correlations. They vanish for all times. Now, as far as we can see, if initial correlations are present but such that they do not introduce any new time scale, the assumptions about the destruction fragment and the expression of the asymptotic correlations are valid within the limits where the collision part of the evolution of ρ_0 can be described by the operator $\Omega\psi$.

In more general situations, for instance, if we irradiate the atom with a light beam, the validity of the kinetic equation should be checked very carefully and we must not expect that it will always be a reasonable description. However, the discussion of the rather simple kinetic limit has enabled us to clarify at least the role of one important contribution. This will certainly be of great help, even in problems where that contribution is not the only one.

Also, one may consider entirely different situations. For instance, we may take atoms in a black-body at finite temperature T. If we start with a situation which is not too far from equilibrium, we may reasonably expect that the system will tend in an irreversible way to the equilibrium distribution, $\rho \sim \exp(-H/kT)$. This implies the existence of an $\mathscr{H}$-theorem, and corresponds to a situation where, clearly, the system has to forget its initial condition. As far as time scales are concerned, we easily notice that the situation is more complex here: we have to take into account absorption and induced emission, which depend on the number of photons present. Also the level shifts depend on the number of photons present.

It is quite exact that the validity of the kinetic equation for long times implies that the long time evolution is dominated by dissipative processes, i.e. processes which give rise to an increase of entropy. As an example, let us consider again the problem of interaction between light and matter. Initially, we assume that the atom is in its *unperturbed* ground state. This state has a higher energy than the exact ground state, as can be easily verified, using ordinary perturbation calculus. So we must expect that some processes will take place which will bring the atom to the true ground state. The situation is not very easy to picture in the bare particle representation but becomes remarkably simple in the physical particle representation. There, initially, we are not dealing with a pure case, but a mixture of the true ground state and the true excited states and one photon. The dissipative processes which then take place may be described as processes of decay of the excited states. Provided we neglect reabsorption of the photons so emitted, we reach a final situation where the atom is in the ground state but where some photons have been emitted. In that sense, we may say that the dressing, once it involves intermediate unstable states, is a dissipative process. Neglecting the possibility of reabsorption (i.e. assuming $t/L^3 \to 0$) it is easy to solve the kinetic equation for this problem and to show that the entropy indeed increases with time.

Since the presentation of this paper great progress has been realized. It will be summarized in a report at the Trieste conference on 'Contemporary Physics' (June 1968, see also forthcoming papers in the Proceedings of the National Academy of Sciences of U.S.A.)

The transformation (5.22) from H_0 to H_R can be obtained *directly* through a transformation of the partition function (1.6) into the form (1.8). The kinetic equation (5.11) remains necessary to study the time evolution, but the meaning of the transformation χ can be studied separately for equilibrium and non-equilibrium. As could be expected (see especially §§ 7–9 of our paper), for non dissipative situations (5.22) reduces to a canonical transformation. But the important point is that it retains a meaning even for dissipative situations. In both cases it leads to a description in which the whole energy is 'in the particles' without any virtual interactions or correlations effects. This point of view is of course very similar to that of 'The Analytic S-Matrix' (see, for example, Eden, Landshoff, Olive and Polkinghorne, *The Analytic S-matrix*, Cambridge University Press, 1966). However, our theory permits us to make a clear distinction between different levels of 'asymptotics'. Indeed, we may deal explicitly with unstable particles *beyond* the statement that they correspond to poles.

General Discussion

I

S. Mandelstam. I would like to make a remark about the point Chew raised concerning the S-matrix in infra-red divergent theories. My remark is purely formal since the problem itself is formal. It has been known since 1937 how to avoid infra-red complications in predicting results of experiments.

Chung has shown that one can calculate an S-matrix in perturbation quantum electrodynamics and that all its elements are finite or zero. The problem is that the S-matrix transforms a state with a finite number of photons into one with an infinite number of soft photons. Since a separable Hilbert space of incoming and outgoing states does not allow an infinite number of photons, we cannot define an S-matrix as an operator in a separable Hilbert space.

I think one can give a satisfactory definition of the S-matrix by making use of some of the ideas recently developed for analysing field theories. Our particular application, however, is done purely within the framework of in and out states and the S-matrix. We consider the operators corresponding to creation and annihilation of incoming and outgoing particles, as well as operators corresponding to the number of particles of any frequency and the operator corresponding to the total energy. We do not consider the operator corresponding to the total number of photons which is not observable.

The Chung S-matrix works in terms of the parametrization of photon states introduced by Glauber. The requirement that a state consist of a finite number of photons, or that it be the limit of a sequence of such states, imposes restrictions on the Glauber function defining it. The S-matrix, when acting on a state satisfying such restrictions, will not produce a state satisfying the restrictions. Let us therefore relax the restrictions in such a way that the S-matrix, when acting on an allowed state, produces an allowed state. The enlarged Hilbert space will no longer be separable. It will consist of the direct sum of a number of Hilbert spaces, each of which forms the basis of an inequivalent representation of the algebra of in and out operators. We shall not require the Hilbert spaces corresponding to all such representations, but the number of representations which we shall require will be at least that of the continuum.

It may appear that we have enlarged the Hilbert space considerably from the Hilbert space from which we started, but we have not made any real physical change. We now have states with an infinite number of soft photons, but, in any experiment with a given accuracy, such a state is indistinguishable from certain states with a finite number of photons. The equivalence may be expressed more mathematically as follows: given any finite set of operators from the in or out algebra and given any state in the enlarged Hilbert space, we can find a state in our original separable Fock space such that the expectation value of the operators in the set for the two states is equal with any pre-assigned degree of accuracy. Thus the Hilbert space has a kind of weak separability with respect to the in or out operators.

We can now define the S-matrix as a unitary operator acting on the enlarged Hilbert space. From this definition one can calculate cross-sections in the usual way.

We make one remark on the analyticity properties of our S-matrix elements. If, for argument's sake, we start with a two-particle state, the final state must contain an infinite number of soft photons if the S-matrix element is to be non-zero. We demand that the total energy of the soft photons be less than some E_0, and let $E_0 \to 0$. We conjecture that the domain of analyticity approaches the usual domain as $E_0 \to 0$.

Finally, we have to consider the question of the analyticity–unitarity approximation scheme. The usual scheme limits the number of particles in the intermediate state; this is clearly inapplicable here. We therefore allow an infinite number of soft photons in the intermediate state, but take the dependence on soft photon momentum to be given by the usual infra-red formula. We have not examined such a scheme in detail, but we do not see any difficulties of principle.

F. E. Low. Could you write a formula for your sequence of states?

S. Mandelstam. Roughly speaking, I consider a sequence such as the following:

$$N(k) = \frac{1}{k} \qquad (k > k_0)$$
$$= 0 \qquad (k < k_0)$$

We then let $k_0 \to 0$. Of course, we also have to specify the relative phases of the photons. This can be done in the Glauber formalism.

N. Cabibbo. Is it not a basic point about these many photon states that actually the important variable is not so much the number of photons as a phase; in other words, these states are coherent: all the photons add up in a classical way?

S. Mandelstam. Yes.

H. P. Dürr. If you have a non-separable Hilbert space of your type which, for example, may be thought of as an infinite direct sum of irreducible Hilbert spaces with different numbers of photons, has one not to worry about the uniqueness of the vacuum in this representation?

S. Mandelstam. No, I do not think so. Only one of the original Hilbert spaces has a vacuum.

N. Cabibbo. What you have to do is to compute the transition from a two-particles state to a state which contains the two final particles plus all the photons up to a certain energy which could be emitted classically. So you have to include not only the photons emitted by the final line but also by the initial line. In fact there are two approaches to the scheme of defining S-matrix elements in this way just by defining them into states which are the particles we want in the end, times an operator which creates the classical amount of photons (it is not all of them, of course, but those up to a certain energy). In a perturbation theory it is well known this matrix element comes out to be finite.

R. Omnès. Do you know how your scattering amplitudes behave under crossing?

S. Mandelstam. If you are worried about the Coulomb phase difficulty, this is a question which, to my knowledge, has not yet been answered. It is, of course, necessary to do so before we have a complete theory.

F. E. Low. What happens to the relation between angular momentum and the asymptotic behaviour? For example, can you, in this scheme, repeat the calculation of the possible Reggeization in perturbation theory of spin-$\frac{1}{2}$ particles and get a finite answer? Do you know anything about whether all relations between angular momentum and asymptotic would be destroyed? The angular momentum–asymptotic relationship is normally for two-particle processes, and you have in fact many particles.

S. Mandelstam. I think this can be assumed at least for a finite number of particles.

II

S. L. Adler. I wish to say a few words about experimental tests of the local current algebra. We have two possible forms of commutation relations that were postulated by Gell-Mann:

Possibility (1): (Integrated algebra)

$$[F_i(t), F_j(t)] = i f_{ijk} F_k(t) \tag{1}$$

$$F_i(t) = \int d^3x \, \mathscr{F}_{i0}(x) \tag{2}$$

Possibility (2): (Local algebra)

$$[\mathscr{F}_{i0}(x), \mathscr{F}_{j0}(y)]_{x_0=y} = i f_{ijk}\mathscr{F}_{k0}(y)\delta(\mathbf{x} - \mathbf{y}) \qquad (3)$$

(and the analogous relations involving axial-vector currents). Possibility (2) is more restrictive than possibility (1) because if there were gradient terms on the right-hand side of Eqn. (3) they would drop out when integrated over all space. The integrated relations have been tested in a number of sum rules. Local relations have only been tested in one sum rule, the Cabibbo–Radicati sum rule. As we shall see, the mechanism of saturation of the Cabibbo–Radicati (C.R.) sum rule is quite different from the mechanism of saturation of more general local sum rules, and it may be that the C.R. sum rule is true, but that more general sum rules following from the local algebra fail.

A simple way to test the local algebra is the following: let us consider the Fourier transform of $\mathscr{F}_{j0}(x)$

$$F_{j,\mathbf{q}} = \int d^3 x e^{i\mathbf{q}\cdot\mathbf{x}} \mathscr{F}_{j0}(x) \qquad (4)$$

then

$$[F_{j,\mathbf{q}}(t), F_{k,-\mathbf{q}}(t)] = i f_{jkl} F_{l,0}(t) = i f_{jkl} F_l(t) \qquad (5)$$

without polynomials in $\mathbf{q}^2$ on the right hand side if the local algebra holds. One sandwiches this commutation relation between proton states and by the familiar methods derives a low energy theorem for the scattering amplitude which describes a proton coming in, current j coming in, a proton going out, and current k going out:

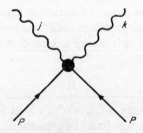

If, in addition, the amplitude obeys an *unsubstracted* dispersion relation in energy, one can express the zero energy value of the amplitude which appears in the low energy theorem as an integral over the imaginary part of the amplitude, which in turn is proportional to the total cross section for (*current*) + $p \rightarrow$ *hadrons*. This gives a sum rule which is a test of the local current algebra.

If you believe in the Regge model, the amplitude involved in the low energy theorem looks like a charge exchange amplitude, in which case its imaginary part is expected to be dominated by ρ-exchange at high energy; if you look at the sum rule you get it turns out that ρ-exchange produces a good enough asymptotic behaviour for the sum rule to be valid.

From the vector commutators, proceeding in the way just outlined, you get sum rules of the form

$$1 = \int_{q^2=q^2}dW[\beta_v^{(-)}(q^2, W) - \beta_v^{(+)}(q^2, W)] \tag{6}$$

where $\beta_v^{(\pm)}$ are related to the cross sections for a fictitious charged photon on a proton going into a state of mass W:

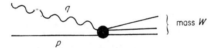

Let us pull out the one nucleon pole, giving

$$1 = (F_1^v(q^2))^2 + q^2(F_2^v(q^2))^2 + \int_{M_N+m_\pi}^{\infty} dW[\beta_v^{(-)} - \beta_v^{(+)}] \tag{7}$$

where $F_{1,2}^v$ are the nucleon isovector form factors. From the axial commutators one obtains

$$1 = g_A^2(q^2) + \int_{M_N+m_\pi}^{\infty} dW[\beta_A^{(-)} - \beta_A^{(+)}] \tag{8}$$

In the inelastic region $\beta_v^{(\pm)} = q^2\bar{\beta}_v^{(\pm)}$, so that going back to the integrated algebra by putting $q^2 = 0$ one gets, in the vector case, the trivial relation $1 = 1$. In the axial case, using PCAC, one gets

$$1 = g_A^2 + \int_{M_N+m_\pi}^{\infty} dW[\sigma_{\pi+p} - \sigma_{\pi-p}] \times \text{known factors} \tag{9}$$

which is the usual g_A sum rule, and is in good agreement with experiment.

In the vector case one can take the derivative at $q^2 = 0$, giving

$$0 = 2F_1^{v\prime}(0) + (F_2^v(0))^2 + \left\{\frac{\partial}{\partial q^2}\int_{M_N+m_\pi}^{\infty} dW[\ ]\right\}_{q^2=0} \tag{10}$$

which is the Cabibbo–Radicati sum rule, and seems to be in agreement with experiment. The C.R. sum rule is a test of the local current algebra.

Another consequence of the vector sum rule which tests the local algebra is the inequality Bjorken derived for electron scattering reactions:

$$\lim_{E_e \to \infty} \left[\frac{d\sigma}{d\Omega}(ep) + \frac{d\sigma}{d\Omega}(en) \right] \geq \frac{1}{2} \frac{d\sigma}{d\Omega}(e'p) \bigg]_{\substack{\text{point} \\ \text{proton}}} \quad (11)$$

where q^2 is the squared momentum transfer between the initial and final electron and E_e is the electron energy

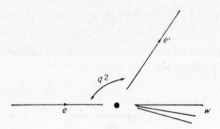

This inequality, and the sum rule [Eqn. (7)] from which it is derived, make an interesting statement. Experimentally, the nucleon electromagnetic form factors behave as

$$F^v_{1,2} \sim \frac{1}{(q^2 + M_v^2)^2} \sim \frac{1}{q^4} \quad (q^2 \text{ large}) \quad (12)$$

so that the elastic contribution to the left-hand side of Eqn. (11) falls off, not as the point proton cross-section (which varies as $1/q^4$), but very much faster, as $1/q^4(F^v_{1,2}(q^2))^2$. However, Eqn. (11) asserts that in the sum over *all* inelastic channels you get something which falls off again as a point nucleon cross-section. Looking at Eqn. (7), we see that the saturation of the C.R. sum rule has not much relevance to the way in which this inequality is satisfied for large q^2, since the C.R. sum rule looks only at the slope at the origin of the curves $(F^v_1(q^2))^2$ and $q^2(F^v_2(q^2))^2$:

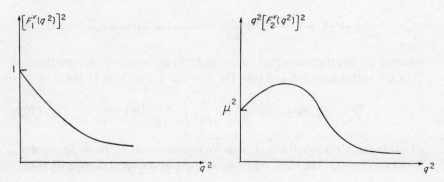

At (and near) $q^2 = 0$, the slopes are opposite in sign and nearly add up to zero, so that *the inelastic continuum is not very important* in the mechanism of saturation of the C.R. sum rule. This cannot be the case for the sum rule in Eqn. (7) or for the Bjorken inequality when q^2 is large, so that we are on the decreasing portion of *both* curves—here the inelastic continuum must be of primary importance.

Recent C.E.A. data on inelastic electron scattering has the following appearance:

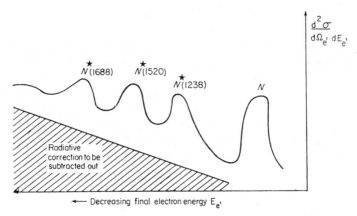

Data is taken with fixed initial electron energy E_e (ranging up to 4.9 Bev) and at a fixed angle $\theta = 31°$ in the laboratory. Only the $N^*(1238)$, $N^*(1250)$ and $N^*(1688)$ are clearly seen. These resonances seem to go down too fast with increasing q^2 to saturate the sum rule of Eqn. (7) or Bjorken's inequality. It would be very premature to say that the sum rule has failed—we may simply have to go to higher resonances than the $N^*(1688)$ to obtain saturation. Of course, as we stated above, if the sum rules do fail, this means that either the no-subtraction assumption is wrong or the local current algebra is wrong.

G. F. Chew. Could you say explicitly for which amplitude the Regge assumption is relevant here?

S. L. Adler. Let us define $A_{\lambda\sigma}$ by

$$\langle p|T(J_\lambda J_\sigma)|p\rangle = A_{\lambda\sigma}$$

Then $q_i q_j A_{ij}$ (protons at rest) is the amplitude for which we need the unsubtracted dispersion relation assumption in the vector case.

F. E. Low. The amplitude which is in the two-photon channel goes from helicity two to helicity anything; in particular with spin 0 target will go to zero helicity, with spin $\frac{1}{2}$ target will go to 0 and 1 helicity, both of which are sense–nonsense amplitudes at $J = 1$.

G. F. Chew. Is there any basis anywhere for being afraid that the asymptotic behaviour would be so bad ...?

S. L. Adler. Only if the Regge model breaks down; in other words, if you accept the Regge model you obtain a unique prediction that the amplitude is ρ dominated, and that the integral over the imaginary part should converge as

$$\int \frac{dv}{v^2 - \alpha} \quad \text{with} \quad \alpha_\rho \sim \tfrac{1}{2}$$

G. F. Chew. Is this the case if you suppose that there is a fixed singularity?

S. L. Adler. A fixed singularity does not contribute to the imaginary part of the amplitude and so does not affect the convergence of the integral.

If you believe that the algebra is correct *and* that unsubtracted dispersion relations are correct, the question which must be answered to interpret experiments is: 'how is Eqn. (7) saturated?' Bjorken has proposed two models for the saturation of Eqn. (7), which I would like to discuss briefly. Both models suggest that the principal inelastic contributions to Eqn. (7) occur for virtual photon energy $q_0 = E_e - E_{e'}$ of order $c + \dfrac{q^2}{2m}$, where c is a constant and m is a mass of order (nucleon mass)/3.

The first model is based on a non-relativistic quantum mechanics sum rule. We write

$$\langle 0| e^{iqx} e^{-iqx} |0\rangle = 1 \tag{13}$$

where $|0\rangle$ is the ground state. Inserting a sum over all intermediate states gives the sum rule

$$\sum_n |\langle 0| e^{iqx} |n\rangle|^2 = 1 \tag{14}$$

If we look at the harmonic oscillator, then this sum rule becomes

$$e^{-q^2/2} \sum_n \frac{(q^2/2)^n}{n!} = 1 \tag{15}$$

Using Stirling's formula, $\dfrac{(q^2/2)^n}{n!} \sim \left(\dfrac{q^2}{n}\right)^n$, we see that the dominant contribution to the sum rule comes from $\bar{n} \sim q^2$, corresponding to a value of q_0 (the difference between the intermediate state energy and the ground state energy) of

$$q_0 = \frac{\bar{n}}{2m_{\text{oscil.}}} \sim \frac{q^2}{2m_{\text{oscil.}}} \tag{16}$$

If baryons are weakly bound quarks, $m_{\text{oscil.}} \sim M_N/3$.

In addition to exciting quark states, a second important mechanism which contributes to the inelastic cross section is the diffraction production of the ρ (or some other 1^- meson):

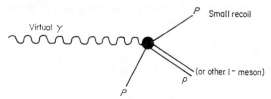

What incident energy of the γ do you need for the momentum transfer to the nucleon to be small enough to get an important diffraction mechanism? The answer is

$$q_0 > \frac{q^2 + m_\rho^2}{2m_\rho} \sim C + \frac{q^2}{2M_N/3} \tag{17}$$

so the second model gives the same estimate as the first. If $q^2 = (1 \text{ Bev}/c)^2$, Bjorken's estimates require that to saturate the sum rule of Eqn. (7) we must excite states with $q_0 \geq 1.5$ Bev, corresponding to an isobaric mass of $W \geq 1.7$ Bev. So the CEA experiments, which go up to $W = 1.7$ Bev, have not necessarily gone high enough. But we see that according to Bjorken's estimates the SLAC machine will certainly reach the region where the sum rule (or electron scattering inequality) should start to get saturated.

Let me conclude by remarking that Bjorken's oscillator model is not as unrealistic as it sounds. Experimentally, the cross section for electro-production of nucleon isobars seems to behave as

$$F(q^2)^2 |\mathbf{q}^*|^{2J+a} \tag{18}$$

where $F(q^2)$ is an elastic electromagnetic form factor, $|\mathbf{q}^*|$ is the virtual photon isobaric-frame momentum, given by

$$|\mathbf{q}^*|^2 = q^2 + \left(\frac{W^2 - M_N^2 - q^2}{2W}\right)^2 \tag{19}$$

where W is the isobar mass, J is the isobar spin and a $(= 1, -1, -3)$ depends on the type of transition involved in the isobar excitation. The sum rule will involve a sum over many isobars:

$$\sum_n C_n F(q^2)^2 |\mathbf{q}^*|^{2J_n + a_n} \tag{20}$$

Eqn. (20) is analogous in form to Eqn. (15): both have a universal factor which decreases with increasing q^2 [$F(q^2)^2$ in Eqn. (20) and $e^{-q^2/2}$ in Eqn. (15)] multiplying a sum over powers of a quantity which grows with increasing

q^2 [$|q^*|^2$ in Eqn. (20), q^2 itself in Eqn. (15)]. Thus the oscillator model suggests that in the actual case the sum rule gets saturated at large q^2 by resonances with spin J_n big enough for the increase of the factor $|q^*|^{2J_n}$ relative to its $q^2 = 0$ value to compensate the decrease of $F(q^2)^2$, giving a product of order unity.

R. Hofstadter. If the quarks in the nucleon are not stationary, the scattered electron spectrum will be spread out and will extend to low energies where the radiative background is very large. It may therefore be difficult to ever test the sum rule.

N. Cabibbo. The problem of radiative tail would be highly reduced in muon scattering experiments.

G. F. Chew. It seems to me that the Bjorken sum rule is physically understandable only if an elementary substructure for the nucleon really exists. Essentially, the statement is that when you hit a nucleon hard enough (with an electron) you get the sum of the cross sections of certain basic constituents.

This kind of relation follows from a completeness relation but without an elementary substructure, particle or field, no meaning is possible for 'completeness'. In a bootstrap regime I would not expect the Bjorken sum rule to be valid.

S. Weinberg. Do you mean that you need something like a quark model to get the local commutation relations?

There are two separate parts to Bjorken's derivation: local commutation relations and the assumption about subtractions. I do not see how either is directly related to the quark model. Local commutation relations are true in a model where the currents are Yang–Mills type fields. The sum rule is not true.

G. F. Chew. Let us take the familiar model of a deuteron as a composite of proton and neutron. If you hit this system with an electron whose momentum transfer is small compared to the reciprocal nucleon size, then the completeness of the wave functions of the deuteron is all you need in order to get the kind of sum rule we are talking about here. The cross section becomes the sum of the neutron and the proton cross sections. But if you hit the deuteron hard enough to start breaking up the nucleons, then you wonder what constituents determine the cross section. Of course if there were a sequence of well-defined substructures you would at a certain level expect, as a reasonable approximation, the sum of cross sections of whatever made up the nucleon at that level. But if you take the strict bootstrap point of view, and if you say that there is no end to this dividing and no fundamental substructure, then I see no reason for expecting high energy cross sections to approach anything simple. It seems to me that the existence of this kind of relationship implies some elementary units that you cannot break up.

S. Mandelstam. I should like to ask to what extent the experimental results are beginning to indicate a breakdown of local commutativity or of the subtraction assumption, rather than a breakdown of the Gell-Mann commutation relations. Do you think that the experimental indications are relative to the possibility of having a finite number of gradient terms?

S. L. Adler. There are two ways (in a local field theory) for the sum rule to break down. (1) Suppose that instead of Eqn. (3) we have a local commutation relation involving a finite number of gradient terms, but that the unsubtracted dispersion relation assumption is still valid. Then we get a modified sum rule of the form:

$$1 + \text{finite polynomial in } q^2 = \int dW [\beta_v^{(-)} - \beta_v^{(+)}]$$

(2) If the unsubtracted relation assumption is wrong there is no reason to expect a finite polynomial subtraction; rather, you would expect a subtraction of the form $1/(q^2 + m^2)$, where m is the mass of some exchanged particle. If you fit experiment accurately enough with a finite polynomial subtraction, then it is more natural to say that the unsubtracted dispersion relation is right and that there is simply some finite number of gradient terms. If, on the other hand, the experiments cannot be explained by a finite polynomial subtraction, then there must be a subtraction in the dispersion relation (or some non-local breakdown of the field theory).

E. C. G. Sudarshan. Let me return to the question of testing current commutation relations. Instead of considering sum rules over infinite ranges which are implied by current commutation relations, we could obtain 'local' relations between amplitudes. Tomozawa derived such a relation for the integrated current commutation relation connecting the scattering length of s-wave pion–nucleon scattering with the axial vector renormalization constant. Since that time, we have seen that these relations are quite general and can be systematically derived using the Ward–Fradkin–Takahashi identities. Last year Balachandran and collaborators at the Institute for Advanced Study showed how these local relations could be used to test the commutation relations of the current densities. Admittedly their computations do have extra model-dependent terms that should be added. It appears to me that these model-dependent tests do supplement the model-dependent tests via sum rules that Adler has discussed earlier today.

R. Hofstadter. I realize that you can probably write sum rules for processes in which you have incoming nucleons and inelastic scattering.

But it seems to me that the things you are looking for are such obvious effects that even in proton–proton collisions you are right to see whether quarks or triplets exist.

S. L. Adler. I have a comment about Professor Hofstadter's remark: I don't believe that there is an analogy between nucleon–nucleon scattering and sum rules. The reason is the difference between a linear unitarity relation and a non-linear unitarity relation the following sense: if you look at the g_A sum-rule, which involves π–N scattering cross sections, what you find is that the $N^*(1238)$ makes a big contribution and that the contributions of the higher resonances get smaller and smaller, which they must do because unitarity bounds them; of course, when spin increases the unitarity bounds increase, but there is a fixed bound which limits how much each resonance can contribute. On the other hand, in the e–N cross sections which come into the sum rules which I have been talking about there is no such unitarity bound, and it may be that even though for large q^2 the cross section for $N^*(1238)$ production is small, the cross section can become large further out in the inelastic region. So I don't know if one can make an analogy between what one sees in the particle scattering case and the case where a weak interaction like electromagnetism (treated only to lowest order) is present.

R. Hofstadter. Is that because of the long range or the short range picture?

S. L. Adler. That's because of the weakness of the coupling constant. The cross-section which appears in the sum rule here is the lowest order in e^2 cross-section only, and not the unitary cross-section, which would involve all orders in e^2. In other words, one is only looking at the one-photon exchange part, for which the unitarity relation is a linear relation and doesn't impose any restriction on how big things can get.

III

S. Weinberg. The comments by Adler and Mandelstam raise questions about the foundations of current algebra, the first by describing experimental results which seem to conflict with the local current commutation relations, and the second showing how some of the successes of current algebra can be re-derived without using commutation rules at all. I feel that it might be useful at this point to try to put the whole of current algebra in perspective, to see which parts are well established and which are more speculative, and to anticipate if we can the directions of future progress.

There is a central body of successful theory which I will call 'classical current algebra'. In brief, this is the use of a symmetry group, chiral

$SU(2) \times SU(2)$, to derive predictions about processes involving soft pions. The strong interaction processes which have been calculated by this method include low energy π–N scattering and $L + N \to 2L + N$ with one pion or all pions soft. Reasonable assumptions about the weak and electromagnetic currents lead to further successful predictions for $\pi \to \mu + \nu$, $K \to 3\pi$, $K \to \pi + e + \nu$, $K \to 2\pi + e + \nu$, and, with some qualifications, $\gamma + N \to \pi + N$ and $Y \to N + \pi$. (One can also go beyond $SU(2) \times SU(2)$ to $SU(3) \times SU(3)$, but it is not yet clear how useful this will be.) There are now three different methods of obtaining the results of classical current algebra:

(a) Current algebra proper is a technique for calculating matrix elements of time-ordered products of currents by using the integrated current commutation relations, the conservation of the vector current and the one-pion dominance of the divergence of the axial-vector current.

(b) An alternative method is to write down a Lagrangian invariant under $SU(2) \times SU(2)$ and then just compute the lowest order diagrams. This works because current algebra ensures that the soft-pion matrix elements are uniquely determined by the properties of the currents, and so any model which embodies these properties must give the right answer. I'll come back to this approach in a minute.

(c) The third method is that suggested here by Mandelstam. I would guess that all the successes of classical current algebra in the area of the strong interactions could be obtained in this way.

To go further we must make additional assumptions. Many paths are open, and we can explore them in various combinations, but if we obtain wrong results this will only invalidate the particular assumptions we have added and not the central body of classical current algebra or its extensions along other paths. I will list some of these paths:

(1) We can make assumptions about the number of subtractions in dispersion relations. This converts some results of current algebra into sum rules, like that for the π–N cross-sections.

(2) We can suppose that the commutation relations used in classical current algebra hold not only for the integrated charges but for the local currents themselves. With reasonable assumptions about the number of subtractions needed this leads to additional sum rules like that for e–N scattering, which was discussed this morning and may be in trouble.

(3) We can try to saturate the commutation relations at infinite momentum or the sum rules derived from them with a few one-particle states, obtaining in this way results that approximate to those which would apply if chirality were a good symmetry of the ordinary sort. There is a related approach, 'partial symmetry,' in which one derives improved results of

this type by writing down Lagrangians invariant under $SU(2) \times SU(2)$ and $SU(4)$, and then taking seriously the way that baryons as well as mesons enter the Langrangian.

(4) We can try various assumptions about the term in the Lagrangian which breaks chiral $SU(2) \times SU(2)$. We know that there is such a term because the pion mass is not zero, and for this reason we cannot count on classical current algebra predictions to better than about 10%, but for the most part it makes little difference just what the symmetry breaking term is. The one exception seems to be the case of π–π scattering. Classical current algebra tells us that $2a_0 - 5a_2$ is about 0.7 pion Compton wavelengths, but we cannot calculate a_0/a_2 without an assumption about how chirality is broken. If we follow the lead of Gell-Mann and Okubo and assume that the symmetry breaking term transforms according to the simplest possible representation of $SU(2) \times SU(2)$, i.e. as the fourth component of a chiral four-vector, then we get $a_0/a_2 = -\frac{7}{2}$ and $a_0 = 0.2$, $a_2 = -0.06$. Mandelstam's approach does not lead to any particular value for the ratio a_0/a_2, so it will be very important to find out experimentally whether or not a_0/a_2 is equal to $-\frac{7}{2}$. If it is, then I think we will have to regard current algebra as arising from a broken symmetry and not from S-matrix theory alone.

G. F. Chew. It seems to me that crossing symmetry should give a definite value to a_0/a_2.

S. Weinberg. I disagree. One can write down an infinite sequence of Lagrangian models in which $SU(2) \times SU(2)$ is broken by a term which transforms like $(N/2, N/2)$, i.e. like a traceless symmetric tensor of rank N, with $N = 1, 2, 3, \ldots$. Each of these models in lowest order perturbation theory gives a π–π scattering matrix element which is manifestly crossing-invariant, but they all give different values for a_0/a_2, namely

$$a_0/a_2 = -\frac{5}{2} \cdot \frac{4 + N(N + 2)}{8 - N(N + 2)}$$

G. F. Chew. Do these models give different p-wave scattering lengths?

S. Weinberg. No, the p-wave π–π scattering length is related by crossing symmetry to $2a_0 - 5a_2$, which is fixed by classical current algebra.

S. L. Adler. Could you clarify the method used for extrapolating the π–π amplitudes? I think some people argued against the method you used.

S. Weinberg. I haven't participated in the argument. Several authors have examined the extrapolation method by working with dynamical models. Some have found self-consistent solutions with large π–π scattering lengths as well as other solutions with small scattering lengths close to those predicted originally, while other authors (most recently

Kang and Akiba) find only the small scattering length solutions. I really don't know who is right. All I can say is that current algebra can predict the π–π scattering lengths if and only if they are small. However, the successes of classical current algebra really show that they *are* small.

S. L. Adler. However, the picture is not completely rosy. Current algebra, PCAC and the neglect of final state pion–pion interactions seem to work well in $K \to 3\pi$ decay, but in $\eta \to 3\pi$ decay something peculiar happens when the same methods are used.

S. Weinberg. There certainly is something funny about $\eta \to 3\pi$. There was a calculation recently by Bardeen *et al.* which tried to show that this decay can be successfully handled by current algebra, but I don't believe it. However, there are other puzzles about η-decay. Probably we will not be able to settle the question of the π–π interaction until the statistics on $K \to 2\pi + e + \nu$ can be improved to the point where the analysis suggested by Cabibbo and Maksymovich becomes possible. Let me move on.

(5) Classical current algebra can be extended by promoting $SU(2) \times SU(2)$ to a gauge invariance of the second kind. This is accomplished either by direct use of gauge invariant Lagrangians, or by assuming that the currents behave like gauge fields and are dominated by the ρ, π and A_1 mesons. Either approach leads to the prediction that the A_1 mass equals the ρ mass times $\sqrt{2}$, which is very well satisfied, except that there may not be any A_1 meson! One also finds relations among the ρ and A_1 widths and the pion charge radius, and there is no doubt that one can also calculate low energy π–ρ, ρ–ρ, ρ–A_1 scattering, etc. In addition, the $\pi^+ - \pi^0$ mass difference can be calculated in the limit $m_\pi \ll m_\rho$, and comes out in excellent agreement with experiment. I feel that these are the results which should cause the greatest discomfort to the S-matrix theorists, because I can't think of any way that Mandelstam or anyone else can derive them without using field theory.

F. E. Low. How do you tell a field from a current?

S. Weinberg. The above results are obtained by using the canonical commutation relations for the gauge fields. We would not expect the usual baryonic currents to satisfy these commutation relations. However, I should mention that Lévy told me the other day that if you work with $SU(3) \times SU(3)$ instead of $SU(2) \times SU(2)$ then these results can be derived not only from the algebra of fields but also from the ordinary algebra of currents. I have not checked this myself, but if Lévy is right then perhaps these results are not so specifically field-theoretic as I thought.

D. Speiser. At which point exactly does $SU(3)$ come in?

S. Weinberg. I wish Lévy were here today to answer that. As I understand it, $SU(3)$ provides enough additional information to eliminate some terms which gave trouble for $SU(2) \times SU(2)$.

R. E. Marshak. Do you say that you cannot derive the spectral-function sum rules without the algebra of fields?

S. Weinberg. You can derive them by making assumptions about the symmetries of current propagators at high energy. However, if you try to derive them from the properties of the $SU(2) \times SU(2)$ currents themselves you must make assumptions which are wrong in every model known, except for the algebra of fields.

In summary, there are three places where experiments could yield insights into the significance of current algebra beyond those provided by the success of classical current algebra. They are: the electron scattering experiments discussed here by Adler, the measurement of the $\pi-\pi$ scattering lengths, and the determination whether or not the A_1 at 1080 MeV is a 1^+ resonance.

I have taken the view that what underlies current algebra is a symmetry group, chiral $SU(2) \times SU(2)$. It may be worth clarifying this a bit. Chiral symmetry doesn't do things other symmetry groups do; it doesn't tell us that the nucleon mass is zero, it doesn't tell us anything at all about nucleon–nucleon scattering—all that it does is tell us about soft pions. This can be made clear in a formalism originated by Schwinger. Consider how the generators X_a ($a = 1, 2, 3$) of the chiral part of $SU(2) \times SU(2)$ act on field operators. We would usually expect relations like

$$[X_a, N] = \gamma_5 \tau_a N$$
$$[X_a, \pi_b] = \delta_{ab}\sigma$$
$$[X_a, \sigma] = -\pi_a$$
etc.

From this point of view we would have to say that chirality is very badly broken, because the nucleon mass is not zero, there is no σ-meson, etc. To avoid this conclusion we might try to do without γ_5's or σ-fields, and instead write

$$[x_a, N] = g_a(\pi)N$$
$$[x_a, \pi_b] = f_{ab}(\pi)$$

The question then becomes: how many such non-linear realizations of $SU(2) \times SU(2)$ are there? I don't know how much there is in the mathematical literature on non-linear realizations of Lie algebras, but I can tell you the answer here: there is essentially just *one* such realization of chiral $SU(2) \times SU(2)$. That is, $g_a(\pi)$ and $f_{ab}(\pi)$ are unique, up to possible redefinitions of the pion field. With $SU(2) \times SU(2)$ realized in this way, it becomes obvious that the chiral invariance of the Lagrangian leads only to relations among processes involving different numbers of soft pions.

I think we can anticipate that the classification and exploitation of the non-linear realizations of general symmetry groups will be an interesting area for future work.

R. E. Marshak. It should be pointed out that the algebra of fields predicts that the current commutator $[j_\mu, j_\nu] = 0$ (μ, ν are space components) in contrast to the quark model. It seems to me that the process $\pi^0 \to 2\gamma$ calculated by Okubo and Cabibbo to test the quark model should also be tested in the algebra of fields model and may give a contradiction (i.e. the prediction will be a very small value for $\pi^0 \to 2\gamma$).

H. P. Dürr. We have heard something about the experimental tests of the current algebra in its integrated and local forms. Could somebody indicate which algebra is tested in the various cases, I mean where we test $SU(2)$, $SU(2) \times SU(2)$, $SU(3)$ and $SU(3) \times SU(3)$? In particular, is there already any definite experimental evidence that the $SU(3)$ current algebra relations are less violated than the mass spectrum?

S. L. Adler. The commutators involving strange mass changing currents (the $SU(3) \times SU(3)$ algebra) are involved in the $K \to \pi e\nu$ and $K \to 2\pi e\nu$ low energy theorem and in the g_A sum rules for hyperon beta decay. There is agreement between theory and experiment, but not nearly to sufficient accuracy to tell whether the strangeness changing current algebra holds exactly, or only up to $SU(3)$ breaking. It will probably not be possible to decide this question in these cases, since PCAC, which introduces $\sim 10\%$ errors, is used. To get a test which does not involve PCAC, one must use a sum rule for $\Delta S \neq 0$ accelerator neutrino cross sections, which are unfortunately very small compared even with $\Delta S = 0$ accelerator neutrino cross sections.

R. E. Marshak. In connection with the successes of current algebra, one should distinguish between the semi-leptonic and non-leptonic applications. The semi-leptonic applications give fairly good agreement allowing for the inaccuracy introduced by using PCAC for the kaon. The non-leptonic applications to $K \to 2\pi$ and the p-wave hyperon decays must still be improved by taking account of the correction caused by $m_\pi \neq 0$.

W. Heisenberg. May I ask whether from the relations which you have written on the blackboard there is any evidence concerning the physical existence of quark particles?

S. Weinberg. I am not sure that I really know. Whether or not quarks exist, it is still an open question how the weak and electromagnetic currents should be constructed.

G. Källén. If one understands a quark model as a model with spin $\frac{1}{2}$ fields fulfilling canonical commutation rules, one gets certain very definite expressions for the right-hand sides of the current commutators (not charge commutators) which are different from the corresponding right hand sides you get, for example, from the model using the Yang–Mills fields. In principle we could distinguish between the different models by looking at these terms. However, I do not know what the exact experimental situation is here or even if any real information is available today.

L. A. Radicati. I would like to ask two questions:

(1) What are the consequences of the third set of the field commutation relations namely those which involve the time derivative of the field? Is there any experimental test of these consequences?

(2) The right-hand side of these commutation relations contains a highly singular term, the product of two operators at the same point. How should we really define it? Would it be possible to modify these commutation relations without abandoning the identification of currents with fields? The change I have in mind could perhaps eliminate the unpleasant quadratic term in the right-hand side.

F. E. Low. The quark model algebra can only be distinguished from other models by testing commutators of space–space components of currents. So far these come in only in the question of the singularity of e.m. radiative corrections to β decay (e.g. $\pi^+ \to \pi^0 + e^+ + \nu$).

R. E. Marshak. This is what I was talking about before with regard to testing the algebra of fields against the algebra of currents. The space–space part of the current commutators in the quark model have a right-hand side which, Okubo has shown, depends on whether there is a fractional charge quark or an integral charge quark, whereas in the algebra of fields the r.h.s. is zero, i.e. the commutator of the space components is zero. So this particular set of commutators distinguishes between the algebra of fields and the algebra of currents, but it can also distinguish between quark models themselves.

R. Brout. Is it possible to go beyond the PCAC type of calculation for namely $\eta \to 3\pi$ decay, presumably $K \to 3\pi$ decay, $\pi^0 \to 2\eta$ etc.? The $\eta \to 3\pi$ decay of electromagnetic has been shown in the first approximation, using PCAC, not to exist.

S. L. Adler. I would prefer to phrase the problem involving η decay in the following way: From PCAC and current algebra, one finds that the $\eta \to 3\pi$ decay amplitude vanishes when any of the pions is extrapolated to zero four-momentum. If you then make the linear matrix element assumption which gave good results in the $K \to 3L$ case, you find that the η decay amplitude vanishes identically. All that this means is that the linear matrix element assumption is incorrect. However, if this linear assumption is incorrect in the η case, why are you allowed to use it in the K case? Perhaps the agreement with experiment in the K meson case is fortuitous.

R. E. Marshak. Two of the three mysteries connected with η decay have apparently disappeared, namely $\eta \to \pi^0 2\gamma$ is not seen and the ratio $\mu \to 3\pi^0/\mu \to \pi^+\pi^- \pi^0$ is close to 1.5.

G. F. Chew. Talking with Heisenberg and Marshak I became aware of a point which I didn't emphasize before. Heisenberg expressed some shock at my interpretation of this point so I will bring it up. As I understand the bootstrap mechanism, it implies that there is no completeness of the Hilbert space in the physical sense. You do, of course, have on mass-shell completeness, but there is no complete set of states which could give a meaning to the sum rules we were talking about before. That is the reason why I don't expect those sum rules to hold which come from local currents, requiring you to go off the mass-shell. Heisenberg pointed out that this is a radical statement because it abandons one of the pillars of quantum mechanics; I agree that it does. The bootstrap mechanism implemented with the S-matrix, as I understand it, abandons the concept of the state vector. There is no state vector that makes any sense in this picture. You work with superposition of asymptotic states, but you have neither completeness nor state vector concepts. I am sorry I didn't make this clear before.

E. P. Wigner. At the 1961 Solvay Congress, Dr. Källén expressed the view that there are physical phenomena which are outside the domain of S-matrix theory. Not only are they not within reach of the bootstrap method, but they are also outside the area of S-matrix theory in its most general meaning. Examples are properties of stationary states of microscopic but particularly macroscopic objects. I am wondering what Dr. Chew's present view is on this subject.

G. F. Chew. I don't think you will obtain the same answer from me as you will from other people, say Froissart and Omnès, who have thought about this. My own view is that the S-matrix is not capable of describing all physical phenomena. In particular it cannot describe macroscopic objects such as solids. Such objects provide the tools which make possible measurement of momentum, but they themselves lie outside the S-matrix. That is my view. I am sure other people do not feel the same way.

D. Speiser. I am not sure whether I have understood what you mean when you talked about superposition and when you said that there is no state vector. You mean the superposition of what?

G. F. Chew. Amplitudes.

D. Speiser. Amplitudes connecting which states?

G. F. Chew. Asymptotic states.

N. Cabibbo. The first of these remarks is connected with the previous discussion. Using the theory by Lee, Zumino, Weinberg and other people, there is a well-defined prediction on the amount of Schwinger terms one has, for example, in the commutator of the fourth component and the space components of the electromagnetic current. This prediction can be tested in $e - \bar{e}$ annihilation into strong interacting particles because the Schwinger term of the electromagnetic current can be expressed as a well defined integral over the annihilation cross section for the $e - \bar{e}$ annihilation to strongly interacting particles, and it has been pointed out recently in a paper whose authors I don't remember (in Phys. Rev. Letters) that the result is very bad and that the cross section is nearly saturated by the ρ alone. If this is true, one wouldn't see much more of the resonances than the one resonance known at present. That particular prediction can be tested and probably will be tested within the next few years.

S. Weinberg. The Schwinger term is known but in terms of an unrenormalized coupling constant.

N. Cabibbo. Oh, it is unrenormalized.

S. Mandelstam. I wonder if Weinberg would like to comment on the calculations by Halpern and Segrè, who claim to have proved from the algebra of fields that certain observable quantities come out infinite.

S. Weinberg. That is a pity.

They seem to come out infinite in almost all models. I wish I knew what to do about it. But are you talking about things that have to do with the fact that the pion mass isn't zero or are you talking about things that are infinite even if the pion mass is zero?

R. E. Marshak. From the algebra of fields, you get an infinite radiative correction even if the pion mass is zero—in contrast to current algebra.

S. Weinberg. Right. But if Mandelstam is correct, there are other infinite radiative corrections which remain infinite even in the limit of the zero pion mass, and that is going to be an interesting problem. Perhaps we just don't know what the commutators are.

G. Källén. I should like to repeat a comment about all calculations of radiative effects (including electromagnetic mass differences) which I made in Rochester about a month ago. In all these calculations you are taking the current commutators very seriously and assume that they hold at least to order e. However, an extra term on the right hand side of order e or e^2 would change all these results without interfering with the successes of the usual current algebra applications.

R. E. Marshak. Has Mandelstam anything to say about the pion mass difference? Can you make any convincing calculations for the electromagnetic mass differences in your approach?

S. Mandelstam. I certainly have not looked much at electromagnetism.

S. L. Adler. I think that one can say something about electromagnetic mass differences in S-matrix theory. Dashen and Frautschi studied this question a number of years ago. They first looked at the non-relativistic Schrödinger equation and derived an S-matrix version of perturbation theory which is equivalent to ordinary perturbation theory and expresses the change in the position of a bound state pole in terms of changes in the complex plane discontinuities of the partial wave scattering amplitude which has the pole. They then applied their method to the relativistic situation and derived an expression for the neutron–proton mass difference by treating the nucleon as a bound state pole in pion–nucleon scattering. There has been controversy over the details of this calculation (removal of infrared divergence, use of linear D functions, etc.) but in any case the method is very interesting.

R. Omnès. The approach to mass differences by Dashen and Frautschi is certainly correct in a relativistic situation since it allows a correct calculation of the Lamb shift. Furthermore the problem of infrared divergences has been solved.

S. Mandelstam. I misunderstood Marshak's question.

L. A. Radicati. Is there a good explanation for the failure of the calculation of the K-mass difference? In particular can it be interpreted by assuming that $SU(3) \times SU(3)$ is broken in a definite way?

F. Low. There are two further reasons why the L calculation should not work so well for the K-mass difference. The first is that $m_k \gg m_\pi$ so that PCAC is necessarily worse for K's than for π's. The second is that whereas the fm_π is given by $I = 2$ amplitude, fm_K is $I = 1$, which Harari has pointed out converges much more slowly at high energy. Therefore the saturation by the first few states is much more doubtful.

IV

SOME RECENT EXPERIMENTAL WORK AT STANFORD UNIVERSITY

R. Hofstadter

Stanford University, Stanford, California

Although electromagnetic interactions were not, for some reason, at the centre of the deep theoretical discussions of our meeting, it is perhaps not completely irrelevant that the only representative of the experimental side here allows himself to present a few aspects of this fundamental and still very active field of research. Of course, we all believe we have a fair understanding of electromagnetic interactions, but I think it is important to realize in concrete and precise terms how well we understand them, and how strict and extensive are the tests to which we submit our theories in this field.

In the short time imparted to this talk I shall limit myself to those aspects which are closest to my own activities, namely the high-energy interactions of electrons and muons with negatons and nuclei. Specifically, I intend to report briefly on some recent experiments which have been done at Stanford and Cornell Universities, at the Stanford Linear Accelerator Center (SLAC), and by groups using the Deutsches Elektronen Synchrotron (DESY) and the Cambridge Electron Accelerator (CEA).

1. Let me begin with a most important test in 'pure quantum electrodynamics', entirely devoted to the celebrated Møller formula. This e^--e^- elastic scattering experiment, conducted by people from Stanford, Princeton and SLAC[1], required 10 years of hard preparation, and was a 'first' in the colliding beams technique. Two beams of 300 MeV electrons from the Stanford storage rings were made to collide, and a spark chamber doublet detected correlated tracks of electrons recoiling with opposite momenta, directly in the centre-of-mass system. In the initial run, 175 hours of counting time allowed them to collect 380 events (after some background subtractions). The corresponding q^2 distribution was compared to the Møller prediction modified by a simple form factor

$$G_\kappa(q^2) = (1 + (q^2/\kappa^2))^{-1} \qquad (1)$$

where $q^2 = 4E^2 \sin^2 \dfrac{\theta}{2}$ is the squared 4-momentum transfer; this G_κ was supposed to combine in a plausible manner any possible electron form factor together with any departure of the photon propagator from the $1/q^2$ law. Of course, the comparison also included the necessary radiative corrections, which become increasingly important at these high energies; for the values of E and θ covered in this experiment, the corrections (as calculated by Tsai[2]) ranged from 4.1–6.4%, but it seems that one may be confident in these estimations.

The absolute cross section could not be measured, but the agreement with the relative q^2 dependence predicted by the Møller formula is good, as is reflected in the lower limit found for κ, namely $\kappa > 0.76$ GeV/c at the 95% confidence level; this also means

$$\kappa^{-1} < 0.26 \text{ fm}. \qquad (2)$$

The experiment has now been repeated at a total centre-of-mass energy of 1.1 GeV, that is, with both rings tuned at 550 MeV. Very recently the Stanford–Princeton group finished collecting about 3000 events; a very preliminary analysis was just completed before I left, and it seems that κ^{-1} will come out smaller than 0.1 fm. About 6000 to 10,000 events are expected in evaluation of the final results.

2. I should like to come now to a rather different subject, which I do not consider to be less fundamental than the preceding one. There are two completely independent ways of making high-precision determinations of the electric charge distribution of nuclei: by using low-energy negative muons from the volume shift of mu-mesic x-rays, and by using electrons from high energy electron–nucleus elastic scattering. In particular, in both cases we can interpret the experimental results partly in terms of the 'mean square radius' of the charge distribution,

$$\langle r^2 \rangle = 4\pi \int_0^\infty \rho(r) r^4 \, dr$$

for spherical nuclei (with $\int d^3r \, \rho(r) = 1$); this number can be converted in a well-known manner either into an equivalent volume shift in mu-mesic transitions with $l = 0$ in the ground state, or into a nuclear form factor in electron–nucleus scattering. More precisely, the assumption is made that in the respective experiments, the 'observed' mean square radius can be interpreted as a sum of two contributions,

$\langle r^2 \rangle = \langle r^2 \rangle_N + \langle r^2 \rangle_{e^-}$ in e–nucleus scattering experiments

$\langle r^2 \rangle = \langle r^2 \rangle_N + \langle r^2 \rangle_{\mu^-}$ in μ–mesonic transitions.

We have now completed at Stanford high-precision electron scattering experiments on the calcium isotopes 40, 42, 44 and 48. The first and last are doubly magic nuclei with almost perfect spherical symmetry; furthermore, the ground state is well separated from the low lying excited levels, so the selection of elastic scattering events is made in good conditions. The analyses of the experiments are made by Ravenhall, Herman and Clark, while the experimental work was carried out at Stanford[3].

Figure 1 shows a comparison of theory (solid lines) with experiment. The dashed lines show the results of a calculation that fitted only the lower energy results (250 MeV and 500 MeV). The 750 MeV results require a modification of the successful parabolic Fermi models for the 250 MeV and 500 MeV results. This slight modification involves putting a small ripple with wavelength 2 Fermis on top of the parabolic Fermi model.

Now, for these same isotopes, the mu-mesic volume shift has also recently been very well measured[4], so a detailed comparison is possible.

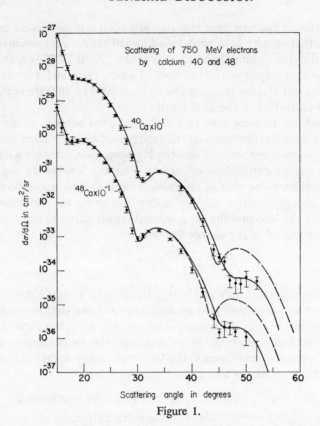

Figure 1.

I may summarize the situation in Table I, where the volume shift of the x-ray transition in ^{42}Ca, ^{44}Ca and ^{48}Ca relative to ^{40}Ca is given (in keV) in the second column as directly obtained by Ehrlich et al.[4], and in the third column as inferred from our scattering measurements. The agreement can be considered to be satisfactory. Note the unexpected change of sign in going from ^{40}Ca to ^{48}Ca. This mu-mesic shift was actually predicted from the earlier electron scattering results.

Furthermore, assuming that $\langle r^2 \rangle_e \sim 0$ from the colliding beam result, we get the limit $(\langle r^2 \rangle_{\mu^-})^{1/2} < 0.2$ fm at the 90% confidence level. Let me mention

TABLE I

Isotopes	Volume shift	e–Ca scattering
40–42	+0.69 ± 0.06	+1.07 ± 0.20
40–44	+0.89 ± 0.05	+0.82 ± 0.10
40–48	−0.47 ± 0.12	−0.41 ± 0.08

another non-trivial result from these calcium measurements, which furnish us with a rather detailed test of the relativistic quantum mechanical theory of scattering as applied to electrons. The elastic scattering cross section on ^{48}Ca was obtained in very high precision runs at 250, 500 and 750 MeV in a wide angular range. Now if you analyze these angular distributions, making an exact partial wave analysis within the framework of the Dirac theory in a central field, completed by an adequate parametrization of the ^{48}Ca electric charge distribution, the 3 independent results produce exactly the same distribution within the rather narrow experimental uncertainties; the fit is as perfect as possible over the full angular range. This also gives us confidence that the charge distribution obtained in such experiments does represent something endowed with physical reality. The agreement between r.m.s. values of radius obtained by the elastic electron scattering method and by the mu-mesic atom studies is also extremely impressive.

3. Let me continue with the same kind of measurements but now applied to the proton. Quite recently the proton magnetic form factor $G_M^{(p)}$ has been determined at SLAC at very high momentum transfers. The DESY groups[5] had already obtained $G_M^{(p)}$ up to $q^2 = 10$ (GeV/c)^2. The SLAC work extends this determination up to 20 (GeV/c)^2. Within its relative accuracy of about 5% (the absolute error is nearly 10%), this experiment fully confirms the 'dipole representation'

$$G_M^{(p)}(q^2)/2.793 = \left(1 + \frac{q^2}{0.71}\right)^{-2} \qquad (3)$$

As already remarked by the CEA group[6] there might exist a small wiggle in the vicinity of $q^2 = 4$ (GeV/c)^2, but this remains to be confirmed; in any case the deviation would not be larger than about 15%. It is now definitely established that the elementary vector meson model

$$G_M^{(p)}(q^2) = \sum_i [\alpha_i(1 + q^2/M_i^2)^{-1} + \beta_i]$$

is unable to represent the data if the summation is restricted to the 3 known 1$^-$ mesons ρ, ω and ϕ. At least one additional ρ' meson is needed, which does not seem to be observed in any experiment. Also, the new data are not compatible with a single exponential decay at high q^2, as predicted by Wu and Yang[7]; they seem to require a superposition of at least two exponentials.

On the other hand, to my knowledge no group has yet reported any disagreement with respect to the straight-line one-photon feature of the Rosenbluth formula; the test has now been extended at SLAC to $q^2 \sim$ 5 (GeV/c)^2. However, it has been remarked[8] that this peculiar behaviour of the Rosenbluth cross-section is not a very stringent test of the one-photon exchange approximation under the present conditions.

4. A more sensitive tool for the detection of a significant two-photon contribution is provided by a comparison at the same q^2 of the high-energy elastic scattering of negatons and positons on protons. The ratio

$$R \equiv \frac{d\sigma(e^+p)/d\Omega}{d\sigma(e^-p)/d\Omega}$$

should equal 1 for pure one-photon exchange. Specifically, if the one-photon exchange amplitude is called $\pm eA^{(1)}$ and the two-photon amplitude $e^2 A^{(2)}$, then essentially

$$R \simeq 1.0 + 4e^2 \frac{\text{Re}\,(A^{(1)}A^{(2)*})}{|A^{(1)}|^2}$$

This ratio had already been measured[9] by the Cornell and DESY groups for q^2 values extending to 1.4 $(\text{GeV}/c)^2$. This has now been pushed at SLAC, by a direct comparison technique, up to $q^2 = 7$ $(\text{GeV}/c)^2$. The problem of the radiative corrections becomes crucial here, as they tend to grow very large and are different for e^-p and e^+p processes[10]. After their inclusion, R is found to be equal to 1 within the $\pm 10\%$ accuracy of the measurement. Apparently the two-photon contribution is not yet detectable at such a high q^2, or we are in trouble with the radiative corrections.

5. I shall now finish by quoting briefly some unrelated but interesting results. In a Stanford–SLAC collaboration we have measured the energy loss of μ^+ and μ^- particles in matter from slightly below 1.0 GeV/c to about 10 GeV/c, with the following arrangement:

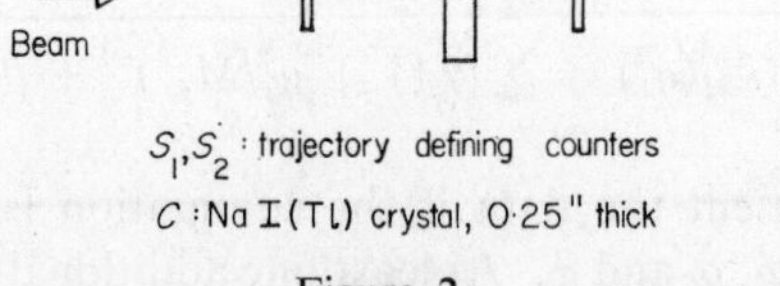

S_1, S_2 : trajectory defining counters
C : NaI(Tl) crystal, 0·25" thick

Figure 2.

The NaI(Tl) crystal could be calibrated carefully with gamma radiations from radioactive substances.

We obtained the following results[11]:
(a) Within the experimental uncertainty of about 2%, the μ^+ and μ^- points are in complete agreement;
(b) the existence of a relativistic rise (from 0.5 to 11 GeV/c) seems to be established beyond doubt, and the observed most probable energy

loss curve is in good agreement with theory. The shape of the Landau straggling curve is also confirmed.

6. In a different arrangement, five NaI(Tl) crystals in tandem were used to search for fractionally charged particles[12]. Such particles could have been produced by pair-production processes induced by photons being produced on a copper target by 12 GeV electrons. The experiment could thus be arranged to provide a convenient way of searching for quarks, with the result that there are apparently no 'miniquarks'. If the mean life of free quarks is assumed to be larger than 10^{-7} sec, then according to calculations by Tsai the experimental results can be interpreted as follows:

$$\text{if} \quad e_q = 0.04e \quad \text{then} \quad m_q > 0.2 \text{ GeV}/c^2$$
$$\text{if} \quad e_q = 0.7e \quad \quad\quad\quad m_q > 1.5 \text{ GeV}/c^2$$

7. Finally, let me also mention that we have succeeded in detecting fissions induced in uranium, tantalum and bismuth by high-energy electron beams[13]. This is a very promising technique, as it benefits from our control of the electromagnetic interactions and also provides a useful complement to the existing studies of fission induced by real photons. This experiment was a joint project of the Thompson group at the Lawrence Radiation Laboratory, University of California.

Thank you very much for your attention.

Acknowledgement.
I appreciate the help of Dr. J. Naisse in the preparation of this paper.

References

(1) W. C. Barber et al., *Phys. Rev. Letters*, **16**, 1127 (1966).
(2) Y. S. Tsai, *Phys. Rev.*, **120**, 269 (1960).
(3) J. B. Bellicard et al., *Phys. Rev. Letters*, **19**, 527 (1967).
(4) R. D. Ehrlich et al., *Phys. Rev. Letters*, **18**, 959 (1967).
(5) W. A. Albrecht et al., *Phys. Rev. Letters*, **18**, 1014 (1967).
(6) M. Goitein et al., *Phys. Rev. Letters*, **18**, 1016 (1967).
(7) T. T. Wu and C. N. Yang, *Phys. Rev.*, **137**, B708 (1965).
(8) S. D. Drell, *Proc. 13th Intern. Conf. on High-Energy Physics*, Berkeley (1966) p. 86.
(9) R. L. Anderson et al., *Phys. Rev. Letters*, **17**, 407 (1966).
   G. Cassiday et al., *Phys. Rev. Letters*, **19**, 1191 (1967).
   W. Bartel et al., *Phys. Rev. Letters*, **25B**, 242 (1967).
(10) Y. S. Tsai, *Phys. Rev.*, **122**, 1898 (1961).
(11) E. H. Bellamy et al., *Phys. Rev.*, **164**, 417 (1967).
(12) E. H. Bellamy et al., 'A Search for Fractionally Charged Particles', *Phys. Rev.*, **166**, 1391 (1968).
(13) H. R. Bowman et al., 'Electron Induced Fission in ^{238}U, ^{209}Bi and ^{181}Ta' (submitted to *Phys. Rev. Letters*).

SOME CONCLUDING REMARKS AND REMINISCENCES

L. Rosenfeld

Our proceedings have now come to an end, and it has become customary that somebody is asked to summarize them. In this case, however, it was felt that a summary would be rather too short, and it was suggested that this might be an opportunity to cast a glance into the past as well, and briefly recall what the previous Solvay meetings devoted to our subject were like.

Allow me to start with a very personal recollection—just to put you into the right mood. When I wrote to Pauli in 1929 to ask him whether I could come to Zürich to work under his guidance, he replied with a very friendly postcard, pointing out that the time was quite favourable, since Heisenberg and he had just started work on the quantization of electrodynamics, 'ein Gebiet', he said, 'das noch nicht ganz abgedroschen ist' (a domain that is not yet threshed out). When I arrived in Zürich, I immediately got a proof of their first paper to read. The first thing I did was to correct a minor mistake in it. I was very proud of it, and wrote a little note which started with the words: 'In ihrer grundlegenden Arbeit, haben Heisenberg und Pauli···'. (In their fundamental paper···). When he saw this, Pauli laughed and remarked: 'Es ist ein ziemlich morscher Grund, den wir da gelegt haben!' (It is a rather swampy fundament we have laid!) This was perhaps a truer statement than Pauli himself thought at the time. I have got the impression that we are still plodding in this mud, some of us trying to dig a little deeper in the hope of finding some firmer ground, others trying to extricate themselves by pulling their own bootstraps. Our situation is still adequately described by a Russian proverb, which Ehrenfest once quoted in a discussion at that time. Russian proverbs are difficult to translate, as we have been told in the United Nations, but they are very picturesque. This one is about a little pig with short legs getting into a muddy pool. When he feels that his nose is too muddy he tries to get it out, but then his tail sinks in, and you perceive that there is no end to his ordeal.

To return to history: shortly afterwards, in 1930, there was a Solvay conference about magnetism. I did not attend it, but being in Brussels at the time, I hovered around, especially at the Club of the Fondation Universitaire, where the participants gathered after the sessions. Most of you do not remember those times of the great depression, and an incident that took place in Detroit, where the streets were then filled with hungry unemployed workers; the president of the Ford Corporation, asked by journalists what he thought about this depression, answered: 'What depression?' Now, I met in the street, near the club, Louis de Broglie, and we got into conversation. I told him that Heisenberg and Pauli had started building up quantum electrodynamics and were encountering great difficulties. With an expression of great surprise he asked: 'What difficulties?'

The club of the Fondation Universitaire was the scene of a famous fight between Bohr and Einstein about the principles of quantum theory. It was the occasion when Einstein thought to have found a counter-example of the uncertainty principle with his well-known box from which a photon is emitted at a certain time, and a weighing of the box before and after the emission determines the energy of the emitted photon. It was quite a shock for Bohr to be faced with this problem; he did not see the solution at once. During the whole evening he was extremely unhappy, going from one to the other and trying to persuade them that it couldn't be true, that it would be the end of physics if Einstein were right; but he couldn't produce any refutation. I shall never forget the vision of the two antagonists leaving the club: Einstein, a tall, majestic figure, walking quietly, with a somewhat ironical smile, and Bohr trotting near him, very excited, ineffectually pleading that if Einstein's device would work, it would mean the end of physics. The next morning came Bohr's triumph and the salvation of physics; Bohr had found the answer that you know: the displacement of the box in the gravitational field used for the weighing would disturb the frequency of the clock governing the photon emission just to the amount needed to satisfy the uncertainty relation between energy and time. We have hardly mentioned gravitation in our discussions, and this may be an indication that none of us feels that the time is ripe for including it into our considerations. Who knows, this may change. It is one of the great open questions whether gravitation does or does not play any part in the interactions at the subnuclear level.

The next occasion when the topic of elementary particles, under this very name, was taken up at a Solvay conference was in 1948. In the intervening period the scope of the subject had been immensely enlarged by the appearance on the scene of the mesons as carriers of the nuclear interactions. At the time of the conference, the last sensation was the production of artificial mesons in Berkeley, about which Serber was supposed to give exciting details. However the technical aids were not then as refined as they are now: there were no microphones, so that Serber could hardly be heard, and the projection apparatus was so bad that it was hardly possible to see the meson tracks that he showed on the screen. This somewhat subdued the expected excitement, and at the banquet concluding the conference, poor Serber was lampooned in a 'meson song', written by Teller and sung to music composed by Frisch for the occasion. Here is the relevant verse:

> 'Some beautiful pictures are thrown on the screen;
> Though the tracks of the mesons can hardly be seen,
> Our desire for knowledge is most deeply stirred,
> When the statements of Serber can never be heard.
>   What, not heard at all?
>   No, not heard at all!
>   Very dimly seen
>   And not heard at all!'

The conclusion of the song gave quite an adequate description of the degree of understanding we had reached at the time:

> 'From mesons all manners of forces you get,
> The infinite part you simply forget,
> The divergence is large, the divergence is small,
> In the meson field quanta there is no sense at all.
>   What, no sense at all?
>   No, no sense at all!
>   Or, if there is some sense,
>   It's exceedingly small.'

From these modest beginnings, quantum field theory was allowed to follow its adventurous course until the Solvay meeting of 1961. There the atmosphere was completely changed through infusion of new blood. The heroes of the day were our friends Chew and Gell-Mann. We then witnessed the birth of the bootstrap idea and the canonization of Regge. There was some alarm when Regge poles were first mentioned, and I remember Professor Wigner's polite and ineffectual efforts to induce somebody to define these Regge poles. However, the discussion went on happily without such a definition. At one point, Chew and Gell-Mann seemed preoccupied with understanding what the old people could have expected from such queer concepts as Lagrangians. Gell-Mann was suggesting: 'They may have thought this and this $\cdots$', and Chew replied: 'No, probably they thought that and that $\cdots$'. It did not occur to them that Heisenberg was sitting there and that it might have been simpler to ask him. What I found most remarkable, however, is that Heisenberg himself did not volunteer any statement about this point.

234 SOME CONCLUDING REMARKS AND REMINISCENCES

Now six years have elapsed, and we are again gathered to survey the same problems. That the lapse of time between successive meetings has narrowed down so much is no doubt an indication of the lively growth of the subject and of its growing interest and relevance. This is due, to be quite honest, not so much to our own efforts than to the brilliant work of our friends who, with one eminent exception, are not here: the experimenters who have furnished us with so much new data to think about. We very much feel the need for more such data to make real progress and to get beyond the stage of our present debates, in which there has been more talk about programmes than achievements. Indeed it was a rather novel situation in physics to be faced not with the question which programme is more successful, but which is more ambitious. Here, I may perhaps turn to Källén for sympathetic support in expressing the hope that this passing stage will be succeeded by one in which a return to some more regular kind of mathematics may give us the firmer ground for which we have been groping so long.

Which mathematics will this be? Will it be C^*-algebra, field algebra, current algebra or some algebra not so current? Perhaps this is a case when we might consult this Persian Sybil, whose wonderful image we admired in Bruges the other day. I noticed, in fact, that Professor Heisenberg was contemplating this lady very intently, but even he did not get any response.

Discussion on the report of M. Gell-Mann*

W. Heisenberg. May I ask a question concerning the idea that the groups (like SU_2, SU_3 etc.) could be derived from the bootstrap mechanism. In nonrelativistic physics one can get a consistent unitary S-matrix for any group structure one might like to connect with the Galilean group. I agree that the Lorentz group (with crossing symmetry, equivalence of forces and particles) gives very strong restrictions. Still it seems difficult to believe that it determines the group structure completely. There may be a solution which is completely symmetrical under SU_2. Still we know that experimentally there is a consistent S-matrix which contains electrodynamics and therefore is not strictly invariant under SU_2.

G. F. Chew. Cutkosky has shown that a simple bootstrap model can quite naturally lead to the presence of a Lie group symmetry in the self-consistent particle spectrum.

G. Källén. May I try to reformulate Heisenberg's question. Quantum electrodynamics, to the best of my knowledge, satisfies the general assumptions about dispersion relations, unitarity etc. which you use in bootstrap, but does not have isotopic spin invariance. However, quantum electrodynamics should not come out of bootstrap if you believe that the result of such calculations corresponds to the world of strongly interacting particles. How is this possible?

R. Omnès. In quantum electrodynamics, the γ is not supposed to be a bound state of $e\bar{e}$ ($Z_3 = 0$ is not imposed as a constraint), as it would have to be in order to fit Cutkosky's argument. Furthermore, even if Z_3 turns out to be zero, the group will not be $SU(2)$ but a one-parameter group.

F. E. Low. I believe the answer to Källén's question is that when Chew used the term analyticity he meant more than analyticity of the scattering amplitude in s and t. Presumably any renormalizable field theory has these properties, at least in a power series expansion in the coupling constant. However it probably does not have analyticity in J, that is, the particles corresponding to the input fields would appear as Kronecker deltas at the appropriate points.

S. Mandelstam. I would like to refer to one question raised by Gell-Mann, namely the ability to define a partially conserved axial current. My remarks refer to the particular questions raised by partial conservation. I shall assume that general currents, e.g. scalar currents or pseudo-vector

* The report of M. Gell-Mann was not available at the time of publication.

currents without the partial conservation restriction, can be defined. The equations for such currents look similar to the equations for electromagnetic currents and I think it would be surprising if one had a solution without the other, but the equations are so complicated that one cannot be sure.

To make things precise, let us consider the approximation where the pion mass is zero; partial conservation then becomes exact conservation. We know that axial current conservation and current commutation relations impose certain restrictions on the strong interactions alone, e.g. certain matrix elements vanish at low energies (Adler self-consistency relations). Another such restriction is that the anti-symmetric part of the scattering of pions against any object is proportional to the isotopic spin of that object. I believe that these restrictions can be proved within the framework of strong interaction theory without introducing currents. One has to make an assumption regarding the conspiracy quantum number of the pion trajectory. Since the question of conpiracies will be discussed further in Chew's talk, I shall not elaborate this further at this stage, except to point out that there are other reasons for believing this assumption to be true.

Having derived the restrictions on strong interactions, I think that one can reverse the arguments by which they are derived from PCAC and current commutations to prove that a conserved axial current can be defined and that the total axial charge satisfies the expected commutation relations.

I cannot say anything about the extra properties implied by commutation relations between charge densities, but the most spectacular predictions of PCAC and current commutators, such as the Adler–Weisberger relation, do not depend on these extra properties.

R. E. Marshak. You mentioned the possibility that the instability of the bootstrap solution might lead to octet dominance for the electromagnetic and weak interactions. It should be pointed out, however, that octet dominance seems to work much better for the weak interactions than for the electromagnetic (e.g. the pion mass difference is large and would have been suppressed by octet dominance) and hence the explanation may have quite a different dynamical origin (e.g. through the soft pion mechanism).

You stated that CP violation may have something to do with orienting the 'strong interaction egg' in SU_3 space. Is not the CP violation effect too small to accomplish this?

M. Gell-Mann. I don't think the size of the effect matters. I mentioned the analogy of the orientation of a magnet in a very weak external field.

To select a direction, you need some perturbation, however small, that distinguishes that direction. If we turn off the electromagnetic and weak interactions and (if different) the Fitch–Cronin interaction, then the bootstrap is unable to select any orientation in $SU(3)$ space. But then the perturbations, no matter how weak, can orient the strong interaction 'egg' when they are turned back on. There is some difficulty, as I have described, in obtaining the observed orientation from electrical perturbations (which distinguish the u quark with charge $\frac{2}{3}$) and weak perturbations (which distinguish the linear combination $s \cos \theta - d \sin \theta$ with $\theta \approx 15°$), since the observed orientation distinguishes the s quark. Thus there is a possibility, however remote, that if a new interaction is required to explain the Fitch–Cronin effect, that new interaction could help in fixing the direction, including the determination of the angle θ.

R. Brout. The orientation of SU_3 under spontaneous breakdown may be partially understood as follows. A Lagrangian which is symmetric under SU_3 is a function of the two Casimir invariants, C_2 and C_3. In the space of the Eightfold Way, C_2 is the usual quadratic form $\sum_{i=1}^{8} \varphi_i^2$ and C_3 is the cubic form $\sum d_{ijk} \varphi_i \varphi_j \varphi_k$. Unlike in SU_2, where there is no C_3, the response to a force in a given direction, say Q, need not be in this direction. In fact, there are two classes of solutions, one in the direction of Q and the other at 120° to Q. The latter class describes a two dimensional manifold, one member of which is the hypercharge Y, if Q is taken to be the principal non-strong force, electromagnetism. It is the latter which is realized in nature. This result is characteristic of the group SU_3 and is the principal reason for considering spontaneous breakdown as the mechanism of mass splitting.

Several problems arise: (1) Clearly a second force is needed to fix the weight plane. Weak interactions would seem not to suffice.

(2) Why is the second class of solution realized and not the first? This is a stability problem and can be answered if one could analyze a term in the mass formula contained in the 27 transforming like YQ. If such a term is important in the sense of minimizing the energy one would then explain the Y type solution.

(3) Finally the 'Goldstone' problem must be explained away. Experimentally one would see a pole in $K \to \pi$ + leptons in the f-form factor of the form $[(m_\kappa^2 - m_\pi^2)/q^2]$ which seems incompatible with experiment. A way out is offered by the bootstrap mechanism.

R. E. Marshak. In considering the question whether the V and A equal time current commutators specify the entire theory, one wonders how to derive the $(V - A)$ weak interaction without introducing the spinor fields and using an argument like chirality invariance of these objects separately. May it not be necessary to add additional commutators with fermion fields?

M. Gell-Mann. I am not sure that fermion 'fields' need to be introduced. A description in terms of currents may be sufficient. In fact, one may get a relatively simple description of hadrons and the strong interactions by expressing the stress–energy–momentum tensor (or gravitational current) $\theta_{\mu\nu}$ for the hadrons in terms of a set of integral spin operators, with well-defined equal time commutation relations that include the V and A currents. Such a description could also be equivalent to the bootstrap description.

This idea has recently been pushed somewhat further by Dashen and Sharp. It amounts to generalizing the notion of a Hamiltonian density, which is θ_{00} expressed in terms of variables with canonical equal-time commutators; here we have θ_{00} in terms of operators with other equal-time commutators, such as those for the vector and axial–vector charge densities. In order to have fermion states as solutions, it is not at all necessary to include fermion fields among the variables. Furthermore, the matrix elements of *any one current* determine the behavior of all fermions as well as bosons.

An interesting point is that if fermion fields *are* introduced and they are quark fields, but no real quarks exist in isolation, then the fermion fields in question are of an unusual type, with no incoming or outgoing parts.

R. E. Marshak. You referred to Low's calculation of the $(K_1^0 - K_2^0)$ mass difference as demonstrating that the strong interaction does not cut off the higher order weak interactions. Can you elaborate on this point since, off-hand, one would expect the strong interactions to bring in a cutoff $\sim m_N$ in this calculation?

M. Gell-Mann. I am only a junior partner in a collaboration with Low and Goldberger on these matters. Francis Low may wish to discuss it at greater length, but I shall just give the gist of the idea.

It has previously been thought that the quadratic divergences in the calculation of the second order weak $K_1^0 - K_2^0$ mass difference might be cut off simply by the strong interaction and require no significant modification of the current–current form of the weak interaction at energies below several hundred BeV. But it appears that the sum rules of current algebra lead to the indestructibility of these quadratic divergences and that probably the formula for the weak interaction requires modification at rather low energies, say a few BeV.

Something similar happens in the case of leptons if we require that neutrino–neutrino scattering at low energies really be much smaller than electron–neutrino scattering. (This is analogous to requiring that the

$K_1^0 - K_2^0$ mass difference be as small as it is.) Of course in the case of leptons the experimental evidence is missing, but the calculations are much cleaner.

F. E. Low. In trying to correct the difficulties due to higher order weak interactions, the cut-off introduced by these higher order effects themselves is enormous (either in a Fermi theory or intermediate vector theory) and much too large to permit the observed $K_1^0 - K_2^0$ mass difference to be as small as it is observed to be; neither can the strong interactions provide a cut-off. Therefore the box is fairly tight, since the empirical cut-off must be several BeV.

S. Weinberg. I would like to take issue with one sentence in Gell-Mann's summary. He said that the chiral symmetry is badly broken. I think this is not necessarily true, and depends on your point of view. You can regard chirality as a symmetry which multiplies baryon fields with matrices like γ_5 and τ, in which case chirality is a good symmetry of the Lagrangian but a bad symmetry of the vacuum, the symmetry breaking being characterized respectively by the pion mass and the nuclear mass. Or you can regard chirality as a good symmetry of everything, but one which when applied to baryon states gives, not a matrix like $\gamma_5\tau$, but one or more soft pions. The two approaches are physically equivalent, but the second suggests that perhaps questions like 'how does PCAC come out of bootstrap?' will be answered when we know how other good symmetries come out of bootstrap. In this connexion, I look forward to having more of Mandelstam's ideas about 'conspiracy conditions' and soft pions.

M. Gell-Mann. We can note two comparisons between the near-conservation of strangeness-changing vector currents and that of axial vector currents. Weinberg emphasizes one of these, namely that for the $|\Delta Y| = 1$ vector currents there is a 'nearby world' in which they are conserved and the masses of $SU(3)$ multiplets are degenerate, while for the axial vector currents there is a 'nearby world' in which they are conserved but *not* by means of degenerate $SU(3) \times SU(3)$ multiplets, rather by means of zero mass pseudoscalar mesons.

I agree with such a remark, of course, and I pointed out another comparison, namely that in considering the purity of representations of $SU(3)$ and $SU(3) \times SU(3)$ at infinite momentum (where independent pion creation by the axial vector charge operators is irrelevant) the representations of the $SU(3)$ algebra of vector charges are rather pure, while there is necessarily very large mixing of the $SU(3) \times SU(3)$ representations.

S. Mandelstam. In reply to Weinberg's question to me, what I say about PCAC is not very different from what has been said before. If one attempts to construct a conserved axial current by solving dispersion relations, one finds poles in the form factors at $q^2 = 0$. The existence of such a current is therefore only consistent if there are particles with $m = 0$, and the coupling constants involving these particles must satisfy certain restrictions. This is simply Goldstone's theorem. I think that the restrictions on the coupling constants can be derived from on-shell dispersion theory without talking about currents. It is then possible to construct a conserved axial current without running into inconsistencies.

I may also remark that the disagreement between Gell-Mann and Weinberg seems to me to be one concerning the meaning of the word 'symmetry'. Let us consider a model with a Lagrangian. Weinberg seems to define a symmetry as a property of the Lagrangian, whereas Gell-Mann defines it as a property of the complete theory. The latter definition is stronger than the former, as it demands that the vacuum be invariant under the symmetry.

Weinberg. No, what I meant was that chirality can be regarded as a good symmetry of the complete theory, provided that we realize it in terms of soft pions.

R. Brout. It is possible to derive at least an approximate test which may distinguish between various models of PCAC. In particular, the Nambu model seems not to stand up under this test. One first observes that in the case of spontaneous breakdown of chirality, the Adler–Weisberger sum rule becomes rigorous. Arguments may be offered that the usual Adler–Weisberger value of $F_A \approx 1.2$ can only change by $0(M^2/\mu^2 \ll 1)$. Here μ is the pion mass and M the nucleon mass. Now as μ gets turned up due to a small bare nucleon mass Nambu quite rightly estimates the change induced in F_A, which contains an infra-red type situation due to the vanishing of the pion mass in the chiral limit. One finds

$$\delta F_A = 0\left(\frac{\mu^2}{M^2} \log \frac{M^2}{\mu^2}\right) \approx 0.3.$$

One sees that F_A in this model calculates to 1.5. The same difficulty in the σ model does not occur. The Adler–Weisberger sum rule is not fulfilled in the Nambu model in the presence of a bare mass, since the matrix elements of $\partial_\mu \mathscr{J}_{\mu 5}$ between physical states do not vanish at infinite momentum transfer. In the σ model, the ability to construct the quantity $\partial_\mu \pi$ permits an unsubtracted dispersion relation for $\partial_\mu \mathscr{J}_{\mu 5}$ matrix elements.

We conclude that the Nambu model is not the correct realization of PCAC, its fundamental discord with experiment being the failure of PDDAC (i.e. an unsubtracted dispersion relation for $\partial_\mu \mathscr{J}_{\mu 5}$ is not true in this model). This is reflected by the large value of F_A which is calculated.

R. Brout. The discussion between Gell-Mann and Weinberg may be reconciled with the remark that chiral symmetry is completely unstable in the presence of a small non chiral perturbation. An infinitesimal mass inevitably mixes the two Lorentz representation of the neutrino with equal weight through the Dirac equation. This is not true in $SU(3)$ breakdown where the states can change gradually in proportion to the amount of dissymmetry.

N. Cabibbo. I would like to clarify my hopes on the question of orienting the symmetry breaking in respect to electromagnetic and weak interactions. The symmetry breaking is characterized by an 8 vector x_i, so that the mass operator is

$$M = M^0 + x_i U_i$$

In the actual world $x_8 = x \neq 0$, $x_1 = \ldots x_7 = 0$, so that the breaking is along the eight-axis of $SU(3)$ space:

$$M = M^0 + x U_8$$

In a bootstrap theory of strong interactions x_i would be determined by a self-consistency equation whose unique form is

$$a(x)x_i + b(x)d_{ijk}x_j x_k = 0$$

The solutions to this equation are determined up to a direction. If we introduce e.m. and weak interactions as driving terms, we get

$$a(x)x_i + b(x)d_{ijk}x_j x_k = c_i^{\text{em}}(x) + c_i^{\text{weak}}(x)$$

This equation has only been explored in the case where the c_i's are independent of x, i.e. neglecting a nonlinear effect of the breaking on the driving force. In this case the solutions have $\theta = 0$ or $\theta = \pi/2$. The hope is that the more general case could give a small value of θ without recourse to other driving interactions, but this possibility has not yet been explored.

E. C. G. Sudarshan. There are two remarkable features of the coupling constants ('charges') for electromagnetic interactions of elementary particles which should be distinguished. The first one, namely that the difference of electric charges between the proton and the neutron is the same as the difference of electric charges between the electron and neutrino, enables charge to be conserved in nuclear beta decay. But there is another feature, namely that the numerical charge of the proton and the electron are equal, which should be also noted. When Gell-Mann talks about the 'direction' of the electric interaction (or the 'direction' of the weak interaction) this second feature is *implicitly* assumed. It comes about as follows: in the space of 'quarks' (or octets) the interactions are second rank tensors which are representable only by ellipsoids, *not* by vectors. The only case in which it makes sense to talk about a direction of a tensor is the one in which this tensor has only one non-zero eigenvalue: the direction of the eigenvector with non-zero eigenvalue is now the 'direction of the interaction'.

There is probably some relation between this feature and the question of the possible factorization of the current operators in terms of field operators. This question is reminiscent of the relation between the Heisenberg and the Schrödinger pictures of quantum mechanics: for the description of pure states with only one non-zero eigenvalue for the density matrix the Schrödinger form is more satisfactory; and the superposition principle has a simple expression in the Schrödinger picture. The special feature of the equality of the electric charge of the proton and electron is suggestive, then, of the desirability of factorization of the currents in terms of field operators.

I would also like to call attention to the possible existence of electromagnetic or weak interactions which are associated with currents with vanishing charge, like the Pauli coupling for electromagnetism. The framework outlined by Gell-Mann ignores this possibility. It would be important to know whether these interactions exist or not.

Finally, in a theory which goes beyond the currents and deals directly with the fields (like the one I present tomorrow), we do have coupling constants; and in such a theory we could talk about the *symmetry of the interaction term* alone. It is for this kind of theory that we imposed chirality invariance to derive our universal V–A weak four-fermion interaction.

S. Fubini. It seems to me that it has become clear that there exist two mechanisms by which nature represents a symmetry. The first way (which applies in first quantized theory) leads to saturation by means of a finite supermultiplet; the second one (which is peculiar of second quantized theory) is connected with creation and absorption of zero mass particles and has led to the beautiful predictions of soft pion theory. The important point is that nature seems to chose the first mechanism for vector currents and the second one for axial currents. Indeed we do not have a decent 'soft scalar particle theory' whereas finite momentum saturation of SU_3 commutators works fine with the lowest multiplet. On the other hand for axial commutators soft pion theory works wonderfully whereas infinite momentum saturation requires a large number of states. I think that this very important difference of behaviour between V and A currents has to be understood before we can say that we have a clear unified theory of the full $SU_3 \times SU_3$ symmetry.

S. L. Adler. Gell-Mann stated that in neutrino reactions one gets tests of the local current algebra without using the PCAC hypothesis. I would like to add a small qualification to this. From current algebra one rigorously gets a low energy theorem for the Compton-like amplitude describing the scattering of a virtual lepton pair on a nucleon [$e\bar{\nu}_e + N \to e\bar{\nu}_e + N$]. This amplitude cannot be directly measured. To get a useful prediction one has to assume that the amplitude obeys an unsubtracted dispersion relation; one can then relate the amplitude at low energy to an integral over inelastic accelerator neutrino cross sections [$\sigma(\nu_e/\bar{\nu}_e + N \to e/\bar{e} + \text{hadrons})$], giving a sum rule testing current algebra in inelastic neutrino reactions. If the sum rule fails, it means that either the local current algebra is wrong or the assumption of an unsubtracted dispersion relation is incorrect (or both).

From the neutrino sum rules, Bjorken has obtained a useful inequality which applies to electron scattering from nucleons,

$$\lim_{E_e \to \infty} \left[\frac{d\sigma}{dq^2}(en) + \frac{d\sigma}{dq^2}(ep) \right] \geq \frac{1}{2} \frac{d\sigma}{dq^2}(ep) \bigg|_{\text{point proton}}$$

q^2 = electron momentum transfer squared

So far, the quantity in square brackets has only been measured for electron energies up to 5 BeV, so that only the contributions of the nucleon and the first few nucleon resonances to $d\sigma(en)/dq^2$ are excited. The elastic nucleon contribution of course is $d\sigma/dq^2$ for a point nucleon × (form factor)² and the first few nucleon resonance contributions fall off *even faster* with q^2, so that for large q^2 the inequality is badly violated. It is too early to decide whether the inequality really fails, or whether one simply has to go to much higher electron energies to excite the states which do satisfy the inequality for large q^2.

INDEX

A1 mass 217
Action-at-a-distance 106
Algebra,
 local 206
 of currents 136
 of observables 2
Analytic functions 114
Analytic structure 143
Analyticity,
 first degree 67–70, 84, 85
 hermitian 69
 maximal 65–67, 74
 of momentum variables 65
 second degree 70, 71, 75, 85, 89
Analyticity–unitarity approximation 204
Angular momentum,
 branch points in 78, 79
Antiferromagnet 13, 14
Antispurion 135
Arbitrary parameters 71
Asymmetry,
 of the groundstate 20, 137
 of the vacuum state 2, 141
Asymptotic behaviour 71, 80
Asymptotic condition 49, 50
Asymptotic equation 173
Asymptotic procedure 156, 161, 192
Asymptotic property 164
Asymptotic states 157, 172, 185, 193
Asymptotic velocity 88
Atom–photon field interaction 157
Axial current,
 conserved 94
 partially conserved 94
 weak 76

Baryon eigenvalue equation 138
Bethe–Saltpeter equation 73, 138, 154
Bethe–Saltpeter model 82
Bjorken inequality 208, 209
Black body 201
Boltzmann approximation 165, 166, 170, 171

Boltzmann distribution of matter 168
Boltzmann entropy 186
Boltzmann representation, 176
 pure states in 177
Bootstrap 66, 67, 70–73, 76, 90, 138, 139, 233
Born approximation 160, 167, 168, 181
Bose–Einstein condensation 12

Cabbibo–Radicati sum rule 206
Canonical commutation rules 89
Canonical distribution, 168, 185
 equilibrium 172
Canonical formalism 105
Causality, 114, 125
 local 129
 macroscopic 68, 87
CDD poles 66
Charge renormalization, *see* Renormalization, charge
Chirality 215
Chirality invariance 54
Chirality symmetry 19
Cluster decomposition property 120, 125, 127
Collisions,
 in moderately dense systems 172
 non-instantaneous 164
Commutation relations,
 local 76
Complex variables,
 analytic functions of 46
Composite particles 89
Conspiracy 83
Conspiracy quantum number M 92
Convex sums,
 method of 119
Correlation effects 172, 175, 177
Correlations, 162, 199
 dynamics of 162
 evolution of 166
 involving two photons 171

INDEX

Coulomb force,
　scalar 20
Coulomb gauge 6
Coupling 135
Coupling constant, 40, 41
　unrenormalized 222
Covariance,
　conflict with locality 105
Covariant field theory 107
CP violation 63
Crossing,
　principle of 65, 69, 90, 205
Crossing relation 91
Cross-section 181
Cross-sections,
　total 77, 78
　and geometric 'radius' 78
Current algebra,
　classical 214
Cut-off methods 35, 44

Daughter sequence 82, 84
Degrees of freedom 122
Density matrix, 161, 182, 198
　evolution of diagonal elements 163, 164
Diagonal representation,
　in quantum optics 123
Diagonalization 110
Diffractive dissociation 77
Dipole ghost 115, 130, 131, 143
Dipole representation 227
Dirac theory 227
Discontinuity equation 69
Dispersion relations, 71
　double 65
　single 65
　unsubtracted 210
Dissipation 183
Dissipative contributions 185
Dissipative effects 178
Dissipative processes 187, 198, 202
Dissipative systems 158, 172, 175, 192
Distribution–theoretic refinements 104
Divergence difficulties 121
Dressing, 172, 202
　in dissipative systems 192

Effective propagator 108
Eigenstates,
　of large S-matrix 110, 112
Eigenvalue problem 33
Einstein theory of transition probabilities 182
Electric current 49
Electrodynamics 70
Electromagnetic interactions 223, 224
Electromagnetism, 55, 70, 142
　induced 56
Electron, 136
　self-stress of 117
Elementary particle physics 1
Elementary particles, 89
　unstable 172
Emission,
　of single photon 170
　of two photons 170
　spontaneous 97, 170
　stimulated 170
Energy levels, 177–179, 185, 186
　effective 158
　influence of lifetime on 185
Entropy, 164, 168, 172, 176, 188, 190, 196
　at equilibrium 186
　Boltzmann 186
Equilibrium,
　approach to 171
Equilibrium distribution function 176
Ergodic states 29
ρ-exchange 207
Excited atoms 172
Excited level,
　spontaneous decay of 200
Excitons 131
External lines 110

Factorizability rule 77
Factorization 79, 80
Fermi liquids,
　Landau's theory of 158
Fermion trajectories 84
Feynman graphs 163
Feynman rules 108
Field theory, 48
　local 89
　nonlocal relativistic 102

Fields,
  redundant 142
  renormalized 186, 190
Fine-structure constant 70, 135
Fixed pole 79, 80
Fock representation 36
Forces,
  long range 20, 134
Forward elastic amplitudes 77
Froissart limit 75, 78, 08
Froissart–Gribov amplitude 81
Functionals,
  of basic fields 38
  of incoming fields 38

Gauge condition 94
Gauge fields 6, 142
Gauge invariances 30
Gell-Mann's local algebra 153
Gell-Mann–Okubo formula 141
Ghost fields 143
Ghost particles 109
Ghost states 188, 190, 196, 197
Gibbs canonical formalism 158, 190
Goldstone boson, 18, 19
  trajectory of 19
Goldstone particles 136
Goldstone theorem 1, 18, 94, 134
Gradient terms 213
Gravitation 232
Gravitational interaction 165
Green's functions 185
Ground state, 184
  asymmetry of 137
Group–theoretical conditions 143
Gupta–Bleuler formalism 125, 127, 142

Haag theorem 192
Hadron electromagnetism 63
Hadron weak interaction 63
Hadronic symmetries,
  external 67
Heisenberg picture 36
Heisenberg's unified theory 107, 115
Hermitian analyticity 69
Hermitian operators 101
Hertz's idea 106
Higher groups 134

Hilbert space,
  pure state in 177
  separable 203
Hypercharge 134, 135

'Improper' representations 188
Indefinite metric, 49, 99, 101, 106, 112, 125
  in Hilbert space 130
Indefinite metric field theory 99
Inelastic electron scattering,
  data on 209
Infra-red divergent theories 203
Inner product 100
Integrated algebra 205
Interactions,
  non-local 48, 90, 133
  strong 55, 60
  weak 56
Internal lines,
  modification of 110
Invariance,
  γ_5 50
  relativistic 102
Isospin group 133

Kinetic equations, 161, 169, 172–178, 180, 197, 199
  validity of 201

Lagrangian field theory, 33, 36, 102
  renormalizable 71
Lagrangians 215, 233
Lamb shift 137, 185
Landau rules 68
Landau singularities 68, 70
Landau theory 178
Large space 102, 114, 121
Lee model 131, 186, 188, 196
Leptons, 70
  conservation of 134
Level shift operator 160
Lifetime correction 160
Lifetimes, 169
  finite 157, 182, 184
  infinite 184
Light and atoms,
  interaction between 200
Linewidth 170

Liouville equation 162
Liouville–von Neumann equation 162, 163, 198
Local algebra 206
Local-in-time theories 105, 106
Locality 102, 127
Long-range forces, *see* Forces, long-range
Lorentz poles 82

Mass renormalization 42
Master equation 129, 139–141, 161
Master-field equation 89
Metric 100
M functions 70
Møller formula 224
Multiparticle production 80
Multiple processes 68
Multiple production 79
Multiple scattering 69
Multi-Regge-pole hypothesis 80
Muon 136

Narrow-resonance approximation 91
N/D method 65, 73
N/D framework,
  finite-channel 87
Negative norm states,
  admixture of 115
Nonanalyzability 112, 115
Non-linear spinor theory 12, 129–144
Non-local interactions, *see* Interactions, non-local
Non-local theory,
  relativistic 113
Nuclear beta decay,
  theory of 53
Nuclear democracy 66, 70, 72, 76
Nuclear physics,
  classical 74, 76
Nucleon electromagnetic form factors 208
Nucleon isobars,
  electroproduction of 211

Off-shell equations 73
One-particle properties 111

Padé approximants 48
Parastatistics 154
Parity 134
Parity doubling 82, 84
Particle interpretation 111
Particle observables 111
Particle states 185
Particle theories,
  relativistic 105
Particles,
  composite 89
  dressed 177
  mathematical 34
  of zero mass 1, 70
  physical 34, 164, 177, 190, 192
  stable 188
  unstable 69, 189, 196, 197
Pauli equation 168
Pauli–Villars regularization method 44, 49, 99
PCAC hypothesis 74, 75, 92, 217
Peak shrinkage,
  asymptotic 77
  elastic 78
  forward 77
Peripheralism 75
Pers model 48
Perturbation field theories 71
Phase space 191, 193
Phase space densities 123
Photon line 137
Photons, 70, 134
  soft 204
Physical observables 102
Physical particles, 34, 164, 177, 190, 192
  and entropy 185
  representation of 196
Physical states 40, 41
Pion, 74–76
  small mass of 74, 76
  trajectory 76
Pion-pole dominance 75
Planck distribution of radiation 168
Plasma,
  interaction with radiation field 171
Polarons 131
Poles 85
Pole–particle correspondence 65, 66

Pole residues, 79
  factorizability of 69
Pomeranchuk trajectory 76–80
Pomeranchuk residues 78
Post-Boltzmannian approximation 166, 170, 171, 178
Power series,
  convergence of 121, 122
Probabilities,
  negative 123
Pure states 191

Quantum electrodynamics 5, 48, 129, 224
Quantum field theory 97
Quantum mechanical states 177
Quantum optics,
  diagonal representation in 123
Quantum states, 178
  and entropy 190
Quarks, 73, 74, 136
  dynamical treatment of 153

Radiation,
  interaction with matter 176, 178
Radiative corrections 224
Raman effect 179
Redundant fields 142
Regge cuts 71
Regge daughters 82
Regge poles 65–95, 233
Regge model 207
Regge trajectories, 71–75, 87, 138
  monotonic 72–74
Reggeized sum rule 71–73, 91, 94
Regularization, 106, 110, 116
  (see also Pauli–Villars regularization method)
Relaxation time 199
Renormalization, 172
  charge 173, 188
  effects 179
  energy and coupling 186
  mass 173
Renormalized fields 186, 190
Residues 75
Resonance light scattering 159, 170, 179, 180

Resonance states 138
Rosenbluth formula 227

Saturation 223
Scalar field,
  neutral 187
Scalar field model, 35
  interaction with point source 35
  renormalized solution 36
Scalar interaction 153
Scattering,
  elastic 170
  inelastic 170
Scattering amplitudes,
  high-energy behaviour of 71
Scattering theory 170
Schrödinger picture 34
Schwinger terms 131, 222
Self-consistency condition 92, 94
Short range 87
Single photon,
  emission or absorption of 167
Singularity,
  fixed 210
Small space 102
S-matrix, 127, 196, 203
  eigenstates of 107, 112, 126
  'large' 108, 126
S-matrix asymptotics 84
S-matrix theory, 65, 231
  equivalence to field theory 73
Specific heats,
  infinite 122
Spectral function 43
Spectral representations 44
Spin 69, 70
Spin relaxation 165
Spinor theory,
  nonlinear 12, 129–144
Spontaneous emission, 169, 170, 180, 181
  Einstein treatment of 161
Spurions 4, 5, 135
States,
  with finite lifetime 187
  stable 193
  unstable 193
Static model 65, 75

INDEX

Statistical description,
  contracted 176
Statistical ensemble 192
Statistical mechanics, 165
  non-equilibrium 161
Statistical mixtures 177, 191
Strange particles 12, 135
Strangeness 134
SU_3 67, 76, 77, 134
SU_3 symmetry,
  broken 141
Substructure,
  elementary 212
Subtraction constants 71
Subtractions,
  number of 215
Sum rules 43
Superconvergence 109
Superconvergence relations 92
Superposition,
  principle of 100
Superselection rule 29, 30
Surface terms 40
Symmetry,
  broken 216
  nonrelativistic 63
  spontaneous breakdown of 131, 134
  secondary,
    of dynamical origin 135
  violations 1

Tamm–Dancoff approximation 132
Thermodynamic properties, 185
  in equilibrium 158
T-matrix 160, 161, 181
Toller analysis 80
Transformation,
  unitary 172
Transition amplitudes 161
Transition probabilities, 157, 161, 179, 181
  Einstein's theory of 182
Transport equation 160
Two photons,
  simultaneous emission or absorption of 179

Ultraviolet divergences 104
Uncertainty principle 123, 197
Unitarity, 90
  analytic continuation of 68
Unitarity condition 197
Unitarity transformation,
  improper 192
Universal Fermi interaction 53
Unphysical cut 87
Unstable particles,
  mass and lifetime of 197
Unstable system,
  decay law of 197

Vacuum,
  asymmetry of 2, 141
  degeneracy of 118, 134, 139
Vacuum expectation values 120
Vacuum quantum numbers 76–80
V–A interaction 53–55
Von Neumann equation 161
Von Neumann–Liouville equation, see
  Liouville–von Neumann equation

Ward identity 19, 142, 143
Wave function, 177
  renormalization of 187
Weak axial current 76
Weak currents,
  locality of 76
Weak interactions, 76, 137
  induced 57
  of strange particles 58
  universality of 94, 142
Weak-interaction effects 70
Weight functions 48, 90
Wigner–Weisskopf model 178, 179, 182

Yukawa interaction 32
Yukawa potentials 90
Yukawa's hypothesis 106

Zero-mass particles 70
'Zeron' 4